TABLE OF CONTENTS

AF890318

SUMMARY

Chapter 1 – The wrong answers

The climate crisis looms, making choices about solutions ever more important. However, the debate on the future is surrounded by hype and vested interests. This report investigates the large-scale technologies that corporations and government are putting forward as solutions to climate change (including carbon capture and storage, hydrogen, agrofuels, and geoengineering), explains why they are unlikely to prevent climate catastrophe, and goes in search of more realistic and socially just solutions.

The international emergency

If a technology can't be developed and deployed in the next decade or so, it's of no use to us as a response to climate change. Waiting for technologies to come on stream before making cuts in emissions will mean we have to make deeper cuts overall.

Chapter 2 – The techno-fix approach to climate change - issues and alternatives

Asking the right questions - the politics of technology

Technology is inherently political. Technologies always interact with power and social structures, usually (though not always) supporting the status quo - corporate power in the global system.

To assess whether or not a technology can be a just and effective climate change mitigation tool we need to ask the right questions: Who owns the technology? Who gains from the technology? Who loses? How sustainable is the technology? Will it be available in time to help?

Problems with techno-fixation: Ignoring the scale and source of the problem

Focusing on technological solutions ignores how the problem of climate change is caused, why it continues to worsen and how much needs to be done to stop it.

The persistent claim that a solution is just around the corner has allowed politicians and corporations to cling to the mantra that tackling climate change will not impact on economic growth.

Climate change cannot be viewed in isolation

Climate change is not the only crisis currently facing the planet. Oil depletion will become a major issue within the coming decade; while competition for land and water, deforestation and destruction of ecosystems, soil fertility depletion and collapse of fisheries are posing increasing problems for food supply and survival in many parts of the world.

Scarcity of investment

Governments spend a limited amount of money on mitigating climate change. Investment in energy R&D is roughly comparable to pre-1973 levels. Where this investment goes is a major issue. Investment needs to go into technologies which can be shown to work in time.

The other parts of the solution - alternatives to techno-fixation

Technological change is part of the solution. But only part. It is useful only as long as it is compatible with, and preferably supports, other changes to the way society works.

Economic change

Growth of the whole global economy means consumption of an ever-increasing amount of goods, using an ever-increasing quantity of energy, mineral, agricultural and forest resources. Replacing the idea of growth as the main objective of the economy would require fundamental changes. Building a new paradigm of economic democracy, based on meeting human needs equitably and sustainably, is as big a challenge as climate change itself, but if human society as we know it is to succeed, the two are inseparable.

Political action

Effective and just solutions to climate change need decision making that involves everyone who is affected by the results of the decision – not just deals between those who stand to profit. If this is to happen, co-operation and maturity will be needed, along with a re-engagement of the population in real politics. The hold of corporate interests over political decisions must be broken.

Social change

To get the changes which will stop climate change and benefit people, we all have to act. Individual action is not enough. We have to find common-interest groups and work collectively, whether as geographical communities or through other connections, to restructure relationships for a low-carbon society.

Chapter 3 – The corporate-dominated approach to solving climate change

The corporate-centred approach to climate change prioritises profits, market based mechanisms and large scale or mass-produced solutions.

In particular, the corporate-dominated approach has led to the establishment of carbon trading schemes, which create profits for polluting companies while failing to encourage, and often damaging, genuinely beneficial projects.

The cleantech boom

'Cleantech' refers to 'any product, service, or process that delivers value using limited or zero non-renewable resources and/or creates significantly less waste than conventional offerings.' It covers four main sectors: energy, transport, materials and water. In 2006, the global clean energy sector alone was valued at $55.4bn (£27.8bn) with cleantech start-ups at 1,500 worldwide. In the reckoning of the cleantech boom there will be winner and losers. The effectiveness and sustainability of the technology are only two of the factors that determine a technology's success, and as far as the market is concerned, not very important ones.

Chapter 4 – The technologies – energy provision

Hydrogen

Hydrogen is a carrier of energy not a source in its own right. A primary energy source – coal, gas or electricity generated from other sources – is required to produce it. Using hydrogen as a vehicle fuel (the main application being considered) would be colossally expensive to introduce, doing so would probably mean a commitment to long-term fossil fuel consumption and, most importantly, producing the hydrogen and compressing or liquefying it to use as a vehicle fuel could have a worse impact on the climate than using petrol.

For hydrogen to be viable as a vehicle fuel it needs numerous technological breakthroughs in all major areas including production, distribution and storage. For hydrogen to reduce greenhouse gas emissions would require a glut of renewable electricity or universal carbon capture and storage within a decade. The likelihood of the infrastructure that would make it work being put in place within four or five decades is slim. The cost of infrastructure to supply just 40% of the USA's light-duty vehicles with hydrogen has been estimated to cost over $500bn (£250bn). So, the technical and economic issues mean it is extremely unlikely to be a climate change solution.

Agrofuels

Agrofuels (also known as biofuels) have been trumpeted as the 'green' solution to cutting oil use for transport, but in fact threaten an environmental and social justice disaster.

Though agrofuels currently supply only a very small percentage of world transport fuel use, their impacts are already significant and the full implications of replacing fossil fuels with agrofuels are beginning to be seen. Many agrofuels increase rather than reduce greenhouse gas emissions. A recent report for the UK Treasury found that when tropical forest is cut down, 'to pay back the initial release of CO2 from clearing the forest would take 60 to 270 years of growing biofuels (using current technologies).' Agrofuels rely on industrial scale agriculture, the growth of which has been dependent on deforestation and cheap fossil fuels. Rainforests are being destroyed in Brazil and Indonesia as agrofuel plantations expand, destroying carbon stores which are vital to regulation of ecosystems. Competition for land between food crops and fuel crops is already contributing to rising global food prices and food riots in poorer countries.

Carbon capture and storage / 'Clean' fossil fuels

Carbon capture and storage (CCS) is a major departure from other climate mitigation strategies. Rather than stopping a damaging activity or replacing fossil fuel use, it allows current activities to continue but captures the carbon emissions and buries them under the ground.

CCS enables the continued dominance and expansion of the fossil fuel industry, particularly coal, with all the inherent unsustainability that goes with increased coal use. CCS is also closely associated with enhanced oil recovery (EOR), where the carbon dioxide pumped into ageing oil wells makes oil available that would otherwise stay in the ground. The technology is unlikely to be proven, scaled up and in widespread use until at least 2030 and possibly 2050, too late to prevent devastating climate change.

Nuclear power

The nuclear industry has jumped on climate change as a last ditch attempt to survive in the face of long-term public opposition, cost escalation and the intractable issue of what to do with nuclear waste. The industry has orchestrated a well-executed spin campaign and has succeeded in putting 'the nuclear option' back on the table. However, nuclear power is not carbon neutral, would not be on-stream in time to plug the energy gap or avoid dangerous climate change, and has huge environmental impacts - not least from the huge quantities of radioactive waste created, which no country has yet developed an effective way of dealing with.

Solar power

Solar energy technologies, including solar panels (photovoltaics) and concentrated solar power, are already functional and are being developed further. There is considerable scope for rapid engineering improvements and for major economies of scale in manufacturing and deployment. Price per kWh is falling rapidly. Technological breakthroughs may be required to significantly increase energy storage potential.

However, solar panels are still very expensive and there are issues with the use of toxic substances in manufacture and depletion of resources such as silver. If all present silver production were devoted to solar, it would supply less than a fifth of the world's electricity.

Solar is one of the emissions-reducing technologies with the greatest potential for scaling-up to provide a significant portion of global energy needs. It also has the advantages of not needing significant new infrastructure and of being usable in isolated areas, both of which make it suitable for Majority World use.

Wind power

Like solar, wind is one of the emissions-reducing technologies with the greatest potential for scaling-up to provide a significant portion of global energy needs, particularly in windy countries like the UK. Since larger turbines and turbines in exposed areas are more efficient, at present wind lends itself to deployment in large wind farms, though efficiency and costs improvements may eventually make small-scale wind viable for more than niche uses. As with solar, however, take-up will be hampered unless better storage technologies can be developed. Overcoming local objections to wind farms may require design improvements (eg to reduce danger to birds) and ownership structures which give more advantage to local residents.

Water power – hydroelectricity, tidal power, wave power

Some large scale water power technologies such as hydroelectric dams and tidal barrages have unacceptably high impacts on local ecosystems, and in the case of large dams, methane emissions have a major climate impact. Other water power technologies such as wave power, tidal stream turbines or tidal lagoons are less developed but potentially more sustainable.

Biomass

Burning biomass is humanity's oldest energy technology – the majority of biomass fuel use globally is still made up of traditional heating and cooking fires and stoves.

The main use for biomass in large scale electricity production is in co-firing biomass with coal in power stations. Co-firing with biomass faces major problems of scalability. For example, to supply 10% of the UK's electricity from biomass co-firing using willow would require plantations covering one-quarter of the UK's land mass.

Biomass does have a role to play where it can be harvested and used sustainably on a smaller scale. The production of gas fuel from agricultural waste including manure (biogas) also shows potential for sustainable expansion, though only to meet a small proportion of total energy demand.

Decentralised energy vs extending the grid

Generating energy through combustion of fossil fuels at centralised power stations and transmission through the National Grid is incredibly wasteful. 61.5% of the energy is lost as heat at the power stations, 3.5% is lost in transmission through the national grid.

One solution is to have local grids producing electricity using a variety of renewable technologies backed up by small power stations which distribute their surplus heat. On the other end of the scale, there is the suggestion of extending electricity grids outwards, connecting up the UK grid with Europe and beyond to harness the enormous potential for concentrated solar power from the Sahara desert.

However micro-generation of electricity by renewables is currently much less efficient than large scale windfarms and district heating networks would require installation of expensive new infrastructure which would generally only be viable in densely populated areas. An extended grid would help to eliminate problems of intermittency of supply as there should always be wind blowing or sun shining somewhere in the extended region, but there are concerns that richer nations may end up effectively colonising the renewable energy resources of the poorer nations.

Chapter 5 – The technologies - geoengineering

The term geoengineering refers to the large scale manipulation of the environment to bring about specific environmental change, particularly to counteract the undesirable side effects of other human activities. Technologies proposed include blasting the stratosphere with sulphates, mirrors in space, covering the deserts in reflective plastic, and dumping iron fertiliser in the oceans.

Reflecting the sun's energy

Some scientists are proposing to increase the amount of solar energy that is reflected back into space. Once any of these schemes is embarked upon it must be maintained for as long as the carbon dioxide emissions that it aimed to counteract remain in the atmosphere (up to 200 years) regardless of any negative impact the scheme is found to have.

Sulphates in the stratosphere

When volcanoes erupt they release sulphates which are known to have a cooling effect on global temperatures by reflecting solar energy back into space. Some scientists are proposing to increase levels of sulphates to simulate this effect. However, the sulphates will have unknown impact on ecosystems, including acid rain and localised climatic impacts potentially causing droughts. Nobel prize winner Paul Crutzen, who advocated research into sulphate aerosols as a last ditch solution to global warming, predicted around half a million deaths as a result of particulate pollution.

Plastic coated deserts

In this plan 67,000 square miles of desert would be coated in shiny plastic each year for 60 years to reflect sunlight. The plastic sheeting would have to be maintained, and periodically replaced, for a century or two. Premature removal would have a rapid global warming effect.

Sunshades in space

This scheme involves a set of 16 trillion transparent, sunlight-refracting shades about 1.5 million km from Earth. The project would require 20 launchers each positioning 800,000 screens every five minutes for ten years and would cost trillions of dollars over 25 years.

Removal of carbon dioxide from the atmosphere

Richard Branson has offered a prize of $25 million for the person who can come up with a commercially viable design to remove large amounts of greenhouse gases from the atmosphere. No remotely affordable technology exists to do it.

Ocean fertilisation

One set of schemes for carbon dioxide capture centres on encouraging the growth of phytoplankton in the oceans, which take up carbon dioxide as they photosynthesise. In theory, some of this carbon dioxide might not return immediately to the carbon cycle. Exactly how much carbon dioxide is sequestered, and for how long, has not been quantified. Ocean scientists, including the IPCC, have warned that this technology is potentially dangerous to ocean ecosystems, unlikely to sequester much carbon dioxide, has the potential to increase levels of other dangerous greenhouse gases such as nitrous oxide and methane, to increase ocean acidification in deep ocean waters, and deplete nutrient loading in surface waters potentially leading to 'dead zones'.

Planting trees

Planting trees to offset carbon emissions has been widely discredited by environmentalists but the sale of these 'carbon offsets' has by no means disappeared. The theory is that through photosynthesis trees absorb carbon from the atmosphere and lock it away. However, this process takes decades and is probably effective only for a few further decades as trees die and rot, releasing the carbon. It in no way compensates for the release of carbon dioxide into the atmosphere from fossil which have been stored carbon for millions of years.

Biochar and Bioenergy with Carbon Storage (BECS)

These two proposals seek to address the issue of removing carbon dioxide from the air at the same time as generating low carbon energy. In BECS, biomass is used to generate heat and power together with carbon capture and storage. The BECS proposal seeks to avoid the onset of abrupt climate change tipping points by quickly converting a large amount of land to biomass fuel production. One figure used is 500 million hectares globally by 2030 - an area larger than the entire European Union. With biochar, where biomass is burnt without oxygen creating biogas and bio-oil can be used to generate heat and power, and carbon rich biochar which is returned to the soil, where experiments suggest carbon could be sequestered for hundreds of years.

Overarching issues with geoengineering

Geoengineering rests on the assumption that humans are masters of the universe and the natural world, and can control and engineer its systems. Climate change has shown that humans do not and probably never will understand the planet's systems well enough to try to artificially engineer a re-balancing of the scales that overconsumption has tipped.

Conclusion

Few deny that humanity now faces a crisis. But the solutions put forward simply don't add up. Technologies are a useful part of the solution, but techno-fixation isn't. Other changes are even more important than technology, and equally technically possible. What is really needed is radical social change that reduces the amount of energy we use to the level that we can sustainably produce, and enables everyone to live decent lives. Whether or not this can be achieved depends on the actions we take now.

CHAPTER I: INTRODUCTION - TECHNOFIXATION

As the the climate crisis looms, choices about solutions become ever more important. However, the debate on the future is surrounded by hype and vested interests. This briefing seeks to assess the large-scale technologies that corporations and government are putting forward as solutions to climate change. It explains why they are unlikely to prevent climate catastrophe, looks at where the decisions about our strategies for survival are being made, and goes in search of more realistic and socially just solutions.

This report includes an overview of the issues surrounding each of the key technologies that are being held up as solutions to climate change, and provides a joined-up analysis and a framework for comparisons. Making the right decisions about technology is vital to avoiding devastating climate change. But many of the technologies being put forward as solutions to this crisis simply won't work, will worsen the situation, cause significant environmental destruction or are not going to be available with a short enough timeframe to help us. Even combined, they would fail to address the whole problem - for example, there can be no big technofix for deforestation, which currently causes around a fifth of all greenhouse gas emissions.

Technofixes are very appealing. They appeal to leaders who want huge projects to put their name to. They appeal to governments in short electoral cycles who don't want to have to face hard choices of changing the direction of development from economic growth to social change. Technofixes appeal to corporations which expect to capture new markets with intellectual property rights and emissions trading. They appeal to advertising-led media obsessed with the next big thing, but too shallow to follow the science. They appeal to a rich-world population trained as consumers of hi-tech gadgets. They appeal to (carbon) accountants: technological emissions reductions are neatly quantifiable, if you write the sum properly. Technofixes appeal, in short, to the powerful, because they offer an opportunity to maintain power and privilege.

But why are they the wrong answers? Surely technology is important? The discourse of 'magic bullets' completely ignores the complexities of different situations and needs, and the widely distributed and poorly measured sources of climate change. In short, it isn't addressing the problem. If we are to have socially just and sustainable solutions to climate change, then we have to all look very critically at how our social and economic systems are failing. If the approach to this problem is primarily technological it has the potential to deepen inequalities between rich and poor as the rich are able to afford access to proprietary technology which enables them to maintain high standards of living while the poor suffer the worst effects of climate change on top of continuing social injustice. This is a recipe for conflict.

The international emergency

Climate change is already happening. Already the air and oceans are warming, growing seasons are shifting, and ice and snow cover have decreased across the world. Extreme weather events such as floods, cyclones and droughts are increasing across the world.[1] The World Health Organization estimates that 150,000 people died in 2000 due to the impacts of climate change.[2]

It's going to get worse. Business as usual means that a temperature rise of around four degrees centigrade above 1990-2000 levels can be expected this century - possibly as much as six degrees.[3]

Many scientists consider that limiting temperature increases to a maximum of two degrees above pre-industrial levels is necessary if we are to avoid devastating climate change.[4] The Intergovernmental Panel on Climate Change (IPCC), which represents the international scientific consensus on the issue, suggests that to have a reasonable chance of limiting warming to two degrees would require a peak in global greenhouse gas emissions by 2015 at the latest, with emissions falling by 50-80% below 2000 levels by 2050 and in particular will need industrialized countries to reduce their emissions by 25% to 40% before 2020, and by 80% to 95% before 2050.[5] This represents a reduction in carbon dioxide emissions of about 5% per year, every year.

It could be worse than this: rising global temperatures will also tip the planet's ecological balance, disrupting ecosystems in ways that provoke production of more greenhouse gas emissions in a feedback effect leading to a catastrophic acceleration of climate change. Two examples of positive feedback illustrate the risk: as soil temperatures rise, soil bacteria respire more, generating more carbon dioxide. As air temperatures rise, tropical forests die back, releasing the carbon they contain to the atmosphere, thereby accelerating the temperature rise. A recent paper estimates that such feedbacks already account for about 18% of global warming.[6]

More than two degrees warming would result in:

- Globally decreasing agricultural productivity, exposing millions more to hunger and leading to increasing social unrest and conflict.
- Up to 3 billion more people at risk of water shortages by 2050, rising sea levels would be destroying coastal cities and farmland, making millions more homeless.
- Overall, one billion people could be climate refugees by 2050.

More than four degrees warming would lead to major changes exceeding the adaptive capacity of many geophysical, biological and socio-economic systems.[7] If we're not on track to sorting out our emissions by 2015, we could be committed to this.

So if a technology cannot be developed and deployed in the next decade or so, it is of little use as a response to climate change. Waiting decades for technologies to come on stream before making cuts in emissions could prove disastrous.[8]

CHAPTER II: THE TECHNO-FIX APPROACH TO CLIMATE CHANGE : ISSUES & ALTERNATIVES

'We can't solve problems by using the same kind of thinking we used when we created them.' Albert Einstein

Asking the right questions

Proposed technological solutions often fail to address the complexities of the problems the world faces because they fail to ask the right questions. Agrofuels are indeed the solution to the transport problem if one asks the very limited question 'how can people run their cars without oil?' rather than the more complex question 'how can people get where they need to go without contributing to climate change?' Answers to the latter question might include limiting the need for travel by relocalising jobs and services, or investment in low-carbon public transport. Asking the right questions in a time of necessary change can lead to solutions which, far from being merely poor substitutes for old ways of doing things, are in fact better alternatives with real social benefits. Positive change can happen in a crisis. But positive change is about much more than technology – a framework is needed to assess the proposed technologies in a systemic context. This section tries to ask some useful questions.

The politics of technology

The debate around technological solutions ranges wider than questioning the risks and benefits of a particular device or technological system. Some of the issues are general, others are specific to technologies proposed as solutions to climate change.

In part, the issue of technology is a question of values, in which the dominant position is currently held by those who might broadly be described as technological optimists. The 'optimist' position maintains that:

- *The general direction of technological development is right and positive (hence 'technological progress').*
- *The drawbacks and risks of technologies are outweighed by the benefits further technological progress will compensate those seen to have lost out in earlier stages of the process and will rectify the problems caused by existing technologies.*
- *Technology can solve social problems.*

The alternative to the optimist position could be called 'technological scepticism'. This approach argues that:

- *The balance between costs and benefits to society from a given technology is often neutral or negative.*
- *The supposed inevitability and rightness of technological progress is a myth.*
- *Social problems require social solutions. The belief that technological solutions can be found to social problems, and to problems caused by earlier technological development, is a dangerous illusion which fails to address the political and social causes of those problems.*

- *Technologies are inherently political and a given technology presupposes and/or encourages a certain structure of control and organisation.*

For example, nuclear power has major political implications since it requires very large-scale developments; a high level of scientific and safety expertise; hierarchical, strictly-controlled organisation and armed security. In turn, a society using nuclear power must continue to have these political or social elements to maintain the technology. Whatever one's opinions of nuclear power as a practical means of electricity generation or a solution to greenhouse gas emissions, it is inherently incompatible with a locally-organised, small-scale, demilitarised economic and social model. This is an extreme example, but one could also look at how mechanised, chemically-aided agriculture presupposes a structure in which farmers have cash for capital inputs, have large enough holdings to make fixed investments worthwhile, and produce primarily for sale rather than home consumption. The introduction of such a technological system into a society based on high levels of self-sufficiency and a non-cash economy where small holdings are cultivated part-time will therefore lead to social change, in which, as with any change, there will be both winners and losers.

It is a mistake to presume that technology is in itself neutral and becomes political only as a result of how it is used and implemented. For one thing, there is no such thing as an abstract, neutral 'technology', only existing, actual 'technologies'. These technologies always interact with power and social structures, usually (though not always) supporting the status quo - corporate power in the global system.

Chapter 3 discusses corporations' interests in technology in more detail – this chapter looks at what considerations might influence assessment of proposed technological developments.

Questions for assessing just and effective climate change mitigation technologies

Who owns the technology?

Not just the hardware (power stations, pipelines) but the patents and other intellectual property. Some technologies in particular – second-generation agrofuels, hydrogen, nano-solar – are likely to be dominated by a few companies owning fundamental patents and charging royalties for their use. How will this affect deployment if these technologies can be made to work? With over four thousand patents on 'clean technologies' granted in 2006 in the USA alone,[9] is it legitimate that possible solutions to climate change be held to ransom?

Who controls the technology?

This is a question of control, and of democracy. If supplies are short, who gets them – those in need, or those who can pay?

Beyond this, who should decide what the solutions to climate change are and which technologies represent the best way forward? How can these decisions be made democratically with participation from the people who will be most affected? Governments make decisions on which technologies to support through public funding. But much more money comes from the private sector, which invests based on potential for profit, not social benefit. And even then, government money often follows the corporate lead – corporations are widely represented on the Research Councils and other bodies which make public funding decisions.

Who gains from the technology? Who loses?

Is the balance of winners and losers just or equitable? For example, agrofuels benefit the companies that grow and trade them. They may keep fuel prices down for vehicle owners, but push up food prices for everyone, and cause land conflicts between plantations and small farmers.

New technologies can also improve social justice: for example deployment of small-scale hydroelectric systems can make reliable, cheap, controllable electricity supplies available to people in areas without a centralised grid.

In most discourse on climate mitigation, economic efficiency is prized above social justice. But promoting new technologies which do not help social justice will entrench and exacerbate existing problems, making them all the harder to deal with in the future. Preferring those new technologies which intrinsically promote equality, democratic control and accessibility has wider benefits than the simple reduction of greenhouse gas emissions.

In relation to climate change, emissions have largely been the fault of the over-consuming rich, while the impacts are being felt most strongly by people in poorer countries. Climate change itself is thus a social justice issue and it is doubly unjust to promote solutions which would worsen the position of those who are already suffering.

Inter-generational justice must also be considered - does a technology impose costs on future generations without conferring any benefits? For example, nuclear waste must be stored indefinitely, long after power stations are closed down; captured and stored carbon dioxide would have to be monitored for centuries after cheap fossil-fuel reserves have been exhausted.

How sustainable is the technology?

Greenhouse gas emissions reductions alone are not sufficient evidence of a technology's benefits. Does the technology deplete other resources, for example by consumption of rare minerals or through its impact on natural ecosystems and biodiversity? Does it have other pollution impacts, such as hazardous waste? Does it encourage or rely on other damaging activities? For example, carbon capture and storage relies on coal mining and encourages greater oil extraction when used for 'enhanced oil recovery'. Can the technology continue to be used in the long term without increasing negative impacts?

What scale of operations can the technology reach?

If a technology is being presented as the answer to a problem, eg a new source of vehicle fuel, it needs to be available at a sufficient scale. So, for example, waste cooking oil is a sustainable source of vehicle fuel, but only available in very small quantities. First-generation agrofuels, even if social justice and sustainability issues could be overcome, could never supply current world vehicle fuel use.

Scalability does not rule out a technology as such, but it is a crucial means of detecting hype around wrong answers which are promoted to allow continuation of business as usual.

When will it be available?

Climate science shows that emissions need to start falling within the next few years, and fall massively in 20 to 30 years. Technologies that are unlikely to be available at an effective scale within that timeframe are not helpful – resources should be diverted from these to more immediately available systems – and to ones which can be proven to work.

The focus of governments and corporations on emissions targets for 2050 can also be viewed as part of a distraction strategy. 2050 is conveniently distant – a target for 2050 allows time to continue business-as-usual in the short term in the expectation of future technological breakthroughs. Tough targets for 2050 are not tough at all. Where are the techno-fix plans for a peak in global emissions by 2015?

Problems with techno-fixation

Ignoring the scale and source of the problem

Focusing on technological solutions ignores how the problem of climate change is caused, why it continues to worsen and how much needs to be done to stop it.

Climate change is the result of over-consumption of fossil fuels and of forest and land resources; about one third of emissions currently come from deforestation and agriculture.[10] This consumption continues to grow in line with economic growth. Technological improvements will not tackle overconsumption or growth in demand; this requires radical changes to economic systems. Without such changes, any technology-based emissions reductions will eventually be eaten up by continued rising demand for energy and consumer goods – efficiency gains will be converted into greater consumption not long-term reduced emissions. Technologies which encourage consumers to maintain high energy use and fossil fuel dependency, such as carbon capture and storage, fail to address unsustainable consumption levels which are the basis of rich country economies and the cause of both climate change and other critical sustainability crises such as declining soil fertility and fresh water supplies.

Even the IPCC now suggests that 85% cuts in global greenhouse gas emissions are needed by 2050,[11] other sources suggest as much as 90% reductions for the UK by 2030.[12] Technology simply cannot deliver these levels of reduction without accompanying changes to demand, which requires economic and social transformation.

Techno-fixation has masked the incompatibility of solving climate change with unlimited economic growth. A rational approach to a certain problem and a set of uncertain solutions might be to say that consumption should be limited to sustainable levels from now, with the possibility of increasing in future when new technologies come on stream. Instead the approach taken has been to continue consuming to the same destructive levels in the expectation that new technologies will come on stream. The persistent claim that a solution is just around the corner has allowed politicians and corporations to cling to the mantra that tackling climate change will not impact on economic growth. In 2005, in his address to the World Economic Forum, Tony Blair said: 'If we put forward, as a solution to climate change, something that would impact on economic growth, *it matters not how justified it is*, it will simply not be agreed to [emphasis added]'.[13] While this view may be slowly changing, it has delayed real action for years.

Climate change cannot be viewed in isolation

Climate change is not the only crisis currently facing the planet. Peak oil (the point at which demand for oil outstrips available supply) is likely to become a major issue within the coming decade; while competition for land and water, deforestation and destruction of ecosystems, soil fertility depletion and collapse of fisheries are already posing increasing problems for food supply and survival in many parts of the world. That's on top of the perpetual issues of equity and social justice.

Technological solutions to climate change generally fail to address most of these issues, except where they may reduce oil use. Yet even without climate change, this systemic environmental and social crisis threatens society, and demands deeper solutions than new technology alone can provide.

Scarcity of investment

Governments spend a limited amount of money on mitigating climate change. Investment in energy R&D (research and development) increased massively in the 1970s as a result of the 1973 OPEC oil embargo, but in the last 30 years R&D investment as a proportion of GDP has continually declined to the point where it is roughly comparable to pre-1973 levels.[14] Where this investment goes is a major issue. While it makes sense to research many options for mitigating climate change, time and resources are limited.

In this context, it is worth looking at the distinction between inventions, or technological breakthroughs, and engineering improvements.

Some proposed technologies rely on things which simply don't exist yet; synthetic microbes which 'eat' carbon dioxide and excrete hydrocarbons; a safe and efficient system for distributing and using hydrogen vehicle fuel; nuclear fusion power. This is not in itself an argument against any investment in these technological possibilities, but it is an argument against reliance on such future technological breakthroughs. Claims that something which doesn't exist yet will solve a known problem, and that it should take most of the available resources, should be viewed simply as a stalling tactic on the part of vested interests.

Other technologies exist, but are benefiting from ongoing improvement; the efficiency and cost-effectiveness of photovoltaic solar panels; devices for exploiting wave and tidal power; energy-efficient electrical appliances. These areas can be relied on to improve, though the timescale may be unpredictable. This is where technology investment needs to focus.

At present, it is the technologies that allow business-as-usual to continue that are receiving the lion's share of investment, regardless of either potential benefit or feasibility. Investment in agrofuels or CCS means less investment in wave power, in decentralised energy or in economic and social changes to limit the need for high energy consumption. The US government is investing $179m (£89m) in agrofuels in 2008.[15] €10bn (£7.9bn) is being spent on an international experimental nuclear fusion reactor in France.[16] Diverting this money away from more immediately practical solutions makes the target of peaking greenhouse gas emissions by 2015 less achievable. It both delays the transition to a low-carbon economy and endangers the future by making devastating climate change more likely.

Transition

Transition – the period of change between the high-emitting societies of today and a distant sustainable future, is a hot topic. But while this change must come, the 'transition' discourse coming from governments and corporations is frequently a cover for arguments that would permit use of technologies in the short term which are known to be unjustifiable in the long term – geoengineering, first generation agrofuels, 'carbon-capture ready' coal fired power stations are argued to be necessary now. But why? Largely to prevent serious change to the rich world's over-consuming lifestyles. The discourse of transition delays the inevitable. When is the real transition to a low-emission, more equitable society even going to start? How long is it going to last?

The other parts of the solution - alternatives to techno-fixation

Technological change is part of the solution. But only part. It is useful only as long as it is compatible with, and preferably supports, other changes to the way society works. Even though these changes are not the focus of this report, a brief summary follows.

Economic change

Current government approaches to climate change consist largely of tinkering with policy and expecting the market to deliver emissions reductions. But the market doesn't want to deliver emissions reductions, it wants to deliver profits. Carbon prices are an arbitrary figure unrelated to the real social and environmental cost of emissions. Meanwhile, policies which may 'harm' the economy have been shied away from. This green capitalist approach is asking the wrong question. Instead of asking how to continue to grow the economy while living on the limited resources left on this planet, it should be asking – why is economic growth seen as more important than survival?

What is growth and do we need it?

The current global economic system is based on the assumption of indefinite growth. While ongoing growth in some areas is possible without more consumption of natural resources and emissions of greenhouse gases, this covers only relatively small sectors of the economy – some services and purely information-based products. Growth of the whole global economy means consumption of an ever-increasing amount of goods, using an ever-increasing quantity of energy, mineral, agricultural and forest resources. Even if energy intensity per unit of economic activity can be reduced, ongoing growth eats up the improvement and overall energy consumption still rises. Renewable energy alone cannot decouple consumption from climate change – just because energy sources are called 'renewable' does not mean there is an infinite amount available that can be accessed sustainably.

Economic growth itself is not a measure of human well-being – it only measures things with an assessed monetary value. It values wants at the same level as needs, and through its tendency to concentrate profit in fewer and fewer hands, leaves billions without the necessities of a decent life. Replacing the idea of growth as the main objective of the economy would require not just changes to government measures and targets, but fundamental changes to financial systems, to the operation of large corporations (also based on the assumption of unlimited growth), and to people's own expectations of progress and success. Building a new paradigm of economic democracy, based on meeting human needs equitably and sustainably, is at least as big a challenge as climate change itself, but if human society as we know it is to succeed, the two are inseparable.

Political action

Politics is about decision making. Effective and just solutions to climate change need decision making that involves everyone who is affected by the results of the decision – not just deals between those who stand to profit. Currently policy making focuses on creating market incentives, which leaves it up to capital to make decisions.

As well as being in thrall to economic growth, current political systems are not equipped to deal with long-term issues. Five-year election cycles and a party system based on petty point-scoring make it almost impossible for politicians to co-ordinate a decades-long process of change. Tinkering with market incentives will not deal with climate change – effort needs to go into long-term planning and real, hard decisions. If this is to happen, co-operation and maturity will be needed, along with a re-engagement of the population in real politics. The hold of corporate interests over political decisions must be broken – privileging profit over sustainability and equity defies democracy and leads to exactly those wrong answers current governments are pursuing.

Social change

Conscious social change means creating systems which allow and encourage people to meet their needs within the limits of the planet's resources. Individual action 'cutting my carbon footprint' is not enough – even if everyone were willing to try, too many things cannot be dealt with as individuals. We have to find common-interest groups and work collectively, whether as geographical communities or through other connections, to restructure relationships for a low-carbon society.

The seeds of this movement are already here, even in the UK. For example, 'Transition Towns' groups are systematically looking at the impacts of their community and creating a programme for co-operating to progressively tackle each one and cut carbon by the timescales required. Communities creating their own sustainability action plans show up what local councils and national government are doing, or failing to do, and to put in place real solutions.

Let's take the transport system as an example. As described in chapter 4, there is no viable techno-fix for the fossil fuels that cars run on (or planes for that matter). The solution has to be to reduce the need for people to travel and goods to be transported. This could be done by making necessary amenities accessible as far as possible within walking distance; commuting less by working less or working closer to home; eating seasonal food suitable for growing locally; necessary transport could be public or pooled. This requires major changes, but they could have positive side effects: revitalising communities, reducing stress, reducing deaths and injuries on the roads.

Wider changes involve reassessing what we value. Consumerism tries to teach us to value possessions, not people, not to think, and to be dependent. Drive, don't walk. Buy ready meals, don't cook, let alone grow your own food. Buy new stuff, don't fix the old stuff (you can't fix it anyway – it's designed so you can't take it apart). Local or organic – the choice is too complicated for our poor little shopper's brains, we're told, so just buy the cheap one. To some extent, we all resist the call to dependency, and we know possessions aren't everything. We could create a culture which encourages this knowledge rather than chips away at it. Increasing the resistance and re-learning practical skills, self-reliance and non-commercial relationships is at the heart of the social change we need to deal with the over-consumption that causes climate change.

CHAPTER III: THE CORPORATE DOMINATED APPROACH TO SOLVING CLIMATE CHANGE

The problems with a technological approach to climate change are in many ways inseparable from the problems with the market-based, corporate-led political approach. Governments have responded to concern about climate change by establishing (inadequate) targets for emissions cuts and incentives for industry to meet them. Because the implementation of solutions is being left to the market, analysing the response of corporations to climate change is key to an assessment of how effective technological solutions will be.

> ***"The way to get the green is to go green.'***
> **Newsweek, March 2007[17]**

From denial to capture - corporations and climate change

Who these days doesn't say they are concerned about climate change? The bizarre situation has arisen where the majority of information about an environmental issue is coming from corporations. Fifteen years ago this would probably have been shocking, particularly for those executives now trying to convince the world of their concern for the planet.

Business strategy on climate change saw a major split in the late 90s. An influential group of companies saw that simply denying that climate change is happening was not going to wash and that more subtle and complex strategies were needed to deal with the threat that tackling climate change would pose to corporate profits.

The denialist propaganda campaign on climate change grew from, and was initially modelled on, the tobacco industry's efforts to discredit evidence of the link between smoking and cancer. The tobacco industry employed 'front groups' to question a range of issues which they classed as 'junk science', including both the cancer issue and climate change.[18] The growing threat of action on climate change led to the creation of the Global Climate Coalition (GCC) predominantly led by oil companies and car manufacturers, to sow doubt about climate science and forestall political efforts to reduce greenhouse-gas emissions.[19] But as the tobacco industry's denialist campaign floundered, with lawsuits taken out against the tobacco giants,[20] so the sheer tide of evidence for climate change undermined the denialist position, and the high-emitting companies began to change tack.

A far better strategy than denial was to be seen to take the lead on the issue, to make sure that, since policy makers are going to have to tackle this issue, they tackle it in a way that is as business-friendly as possible. In 1997 major companies began defecting from the GCC, making grand public statements about how they had seen the light on climate change and were going green.[21] Oil companies rebranded themselves as 'energy' companies and made high-profile purchases in wind and solar energy. This was heralded by many NGOs as a major step forward.

Climate denial is now basically dead in the water, with the USA as its last major stronghold. However, it was an effective strategy – if one recognises that the point of denial was not to win the debate but to make it look as though a debate existed, and thereby delay the inevitability of governments taking climate change seriously. By playing themselves off against the demon deniers, the more positive, 'capitalism is the solution', rebranded companies have been very successful. So the twin strategy of denial and false solutions may in fact have worked better for industry than an either/or approach.
The third strategy for delay is distraction, particularly using new technologies. Companies may invest in a technology as a PR exercise without it having any real prospects for solving the problem. This can be seen as buying time to allow destructive sectors such as the oil, coal, aviation and automobile

industries to continue to make big profits as people wait for the technological 'alternatives' to arrive. In effect, companies are saying to the public: 'Don't worry about climate change, we've got it covered. We may not have everything right yet but trust us we're getting there.' This is the underlying message of every corporate sustainability report from BAT to General Electric. Technologies are being used for spin by governments and corporations - the independent of sources of information are crucial.

Alongside the staged public debate, corporations' influence on government has given them a powerful voice in the development of climate change policies. With the support of the corporate elite, Al Gore led the US team in the negotiations for what became the Kyoto Protocol, and achieved a massive victory for business interests by introducing emissions trading mechanisms. This meant that polluters had license to continue to pollute (see below).[22]

What do corporations want?

Profits

Corporations are legally obliged to put profit above all other considerations.[23] Any seemingly positive actions will always be the ones the companies think will be in the best interest of their profit margins. Agrofuels companies talk about climate change; Unilever, the food and 'personal care' giant, faces rising prices for raw materials, particularly palm oil, as a result of agrofuel demand and has thus published a report warning of the environmental dangers of agrofuels.[24] Companies' concerns about climate change are motivated by self-interest. Proof of this can be found in the investments these companies make. BP invests in solar but is still busy building new oil pipelines. The same oil companies that say they have the problem in hand are pouring money not only into new oilfields but into tar-sands and shale oil – fossil fuel sources with even higher emissions than conventional crude oil.

It is permitted for company directors to act based on protecting future profits, which could include working to prevent a major crisis in the society in which the company operates. But to do this effectively would require collaboration between many companies: without collaboration, the leaders in scaling back emissions would see their short-term profitability damaged and be savaged by the stock markets. Collaboration between large numbers of companies for anything other than immediate interests (eg lobbying against restrictive legislation) is almost unknown at present. Moreover, dealing with climate change would inevitably be the death-knell for some industries – these companies would be unlikely to sign up.

Even if collaboration were to take place, it would still be predicated on corporations' long-term interests, which are not the same as the long-term interests of society.

Market based mechanisms

Corporations want to avoid any regulation that would impact on their ability to make profits. Market based mechanisms, where companies are given financial incentives to encourage them to cut down emissions through tax breaks or emissions trading schemes, are more acceptable to most companies if the alternative is regulation. Market based mechanisms work on the logic that corporations will work in the best interest of society if only the rewards are right. (See section on carbon trading.)

Predictability

Second only to profits, corporations want a predictable business environment. In relation to climate policy, this means knowledge of what taxes on pollution will be charged, what subsidies for energy efficiency will be paid and what public investment in new technologies will be available, as well as knowing whether government policy is likely to support or oppose particular infrastructure developments such as new coal-fired power stations, airports or wind farms. Corporations can adapt to a more restrictive environment (though they will lobby very hard to avoid it) but only if they know that environment will be reliable years into the future.

Competitive advantage

By investing ahead of the rest of the field in new technologies and pushing for favourable market based mechanisms, companies hope to be able to gain advantage over their competitors. They can also reap rewards from the PR benefits associated with being seen to be in the vanguard.

Focus on lifestyles

It suits corporations and government to portray the climate change problem as one of consumer choices rather than economic structures. This shifts responsibility away from producers and legislators while ignoring government's role as representatives defending the public interest and the corporate role of stimulating desires and over-consumption. Consumption becomes the solution, not the problem.

Large-scale, mass-produced, ongoing solutions

Some products and markets are more suited to domination by large corporations than others because of economies of scale, profit margins, labour-to-capital ratios, opportunities for standardisation (or lack of them), and various other factors. Thus, for example, the oil trade is dominated by huge transnational corporations, while hairdressing is almost entirely the preserve of very small businesses.

In the growing markets for emissions-reduction technologies, it can be seen how the large-scale, centralised technologies (nuclear power, dams, tidal barrages, coal-fired power stations with carbon capture and storage), or mass-produced technologies which create an ongoing market by selling a quickly consumed product (agrofuels, hydrogen) are achieving most of the hype and investment, while technologies which require smaller, more individualised installation and are likely to involve an end product which is a one-off purchase of equipment (river-run hydroelectricity, farm-scale biogas production, solar thermal) are relatively neglected, despite having greater emissions reductions relative to the energy they produce. The classic renewables technologies, wind and photovoltaic solar panels, fall somewhere between the extremes. The big corporations are concentrating where the profits are, and avoiding the markets where smaller firms would have an advantage.

Carbon trading - how it works and its effect on new technologies

Why carbon trading is important for this report

Carbon trading is not a technology, it is a mechanism promoted by governments and corporations for encouraging emissions reductions. At present, it is the dominant system for meeting emissions targets. As such, it has an important influence over which technologies are taken up and how technological innovation occurs.

Carbon trading is also of interest as it demonstrates how far corporate involvement in negotiations on climate change treaties has created a corporate-friendly but unworkable system based entirely on free-market ideology.

What is carbon trading?

Carbon trading attempts to tackle greenhouse gas emissions by turning the use of the Earth's carbon cycling capacity into a commodity which can be valued and exchanged in a market. A limit (or cap) is placed on the amount of greenhouse gases that can be emitted, and this amount is divided up among those wishing to pollute. Carbon trading is the key mechanism introduced through the Kyoto Protocol, by which industrialised countries (known in the treaty as Annex 1 countries) seek to meet their emissions reduction targets. In Europe emissions are traded through the European Emissions Trading Scheme (EU ETS).

The type of emissions trading scheme introduced by the Kyoto Protocol is a hybrid system which allows polluters who have emitted more than their quota to:

> *1. Buy rights to pollute from polluters that have emitted less than their quota - this is called cap and trade.*
>
> *2. Buy credits from poorer, non-Annex 1 countries. Credits are created through 'carbon saving' projects (such as factory efficiency improvements, tree plantations, or renewable energy projects) - this is called baseline and credit.*

The Kyoto Protocol introduced two of these baseline-and-credit schemes: the Clean Development Mechanism (CDM) for non-Annex 1 countries (eg projects in the South) and Joint Implementation (JI) for projects in other Annex 1 countries (eg projects in Eastern European countries).

The cost of emissions permits – the carbon price – is negotiated by the market. At the time of writing, the EU current carbon price hovers around €22-24/tonne carbon dioxide. In theory, carbon trading should result in a set level of emissions cuts being made in the cheapest way possible.

The way quotas are divided up has been based largely on how much countries emit already. Countries have negotiated national quotas as part of the Kyoto process and have passed on a portion of their quota to heavy industry and energy sectors free of charge. In the UK 46% of the national quota was handed out to around 1,000 installations, amounting to a total asset transfer of €4 billion (£3.2bn).

Main actors

Kyoto Annex 1 countries:

Australia, Austria, Belarus, Belgium, Bulgaria, Canada, Croatia, Czech Republic, Denmark, Estonia, Finland, France, Germany, Greece, Hungary, Iceland, Ireland, Italy, Japan, Latvia, Liechtenstein, Lithuania, Luxembourg, Monaco, Netherlands, New Zealand, Norway, Poland, Portugal, Romania, Russian Federation, Slovakia, Slovenia, Spain, Sweden, Switzerland, Turkey, Ukraine, United Kingdom.

Countries selling most CDM credits:

China, Brazil, Korea and India (together totalling 86% of CDM credits)

Carbon Exchanges, Brokers, Traders:

European Climate Exchange (ECX), Chicago Climate Exchange (CCX), Ecosecurities, Point Carbon, Climex, SendeCO2,

CDM and JI Credit Verifiers:

Det Norske Veritas, TÜV Süddeutschland, SGS Société Générale de Surveillance, Japan Quality Assurance

Lobby groups:

Environmental Markets Association, International Emissions Trading Association, International Petroleum Industry Environmental Conservation Association, Transatlantic Business Dialogue,

Companies in receipt of largest numbers of free credits from UK government:

Corus, E.ON, RWE, EDF, Keadby Generation (owners of Fiddlers Ferry and other power stations), Drax, Scottish Power, Shell, BP, Exxon Mobil, Centrica, British Energy, Blue Circle (owners of Lafarge Cement), Total, Scottish and Southern Energy

NGOs:

World Resources Institute, Environmental Defence, WWF

Extreme complexity makes it unworkable

Carbon trading critic Larry Lohmann points out that 'the Kyoto Protocol's market has set up one of the most centralised, opaque, complicated and jargon-ridden international processes ever seen, while the EU ETS is perhaps the most complex, impenetrable piece of environmental legislation Europe has ever known.'[25] Emissions trading advocates point to the success of emissions trading in dealing with sulphur dioxide emissions in the USA. But emissions trading has been tested only in one country, with limited success and under very different conditions.[26]

For an emissions trading scheme to work there needs to be adequate monitoring of emissions, enforcement of penalties and a limited number of sources of emissions. These conditions do not exist with carbon trading. The level of greenhouse gases countries emit cannot be precisely quantified in the way that is necessary for the system to really work. One study puts the level of uncertainty at 4-21%,[27] another 10-30%.[28] The UK Environment Agency states that 40% of sites do not have satisfactory monitoring of emissions.[29] In addition there are massively more sources of emissions for carbon dioxide than there are for sulphur dioxide. John Henry, a businessman with successful experience in brokering US sulphur dioxide emissions permits, said that - given the lack of ability to monitor so many source points and the absence of a national regulatory enforcement mechanism – international carbon trading will 'give the mechanism of emissions trading a bad name.'[30]

These are just the problems with cap-and-trade. Baseline-and-credit trading is even more complicated – accurately verifying emissions from projects is almost impossible. It is in the interest of all parties to exaggerate the emissions saved. If baseline-and-credit projects are to compensate for increased emissions in industrialised countries, they need to save more carbon than would have been emitted otherwise if the carbon credit money had not been available to the project (the 'baseline' emissions scenario). So the project has to demonstrate that if it did not go ahead there would be more emissions, and that the project can only go ahead because carbon credit money is available.

In reality, any 'baseline' scenario that a company puts forward will be full of assumptions that cannot be proven and any variation in the assumptions will give wildly different figures of how much carbon is saved. There is a strong incentive for projects to inflate the level of emissions in the 'baseline' scenario so that they gain more credits. There is also a strong incentive for companies to apply for carbon credits for projects that would have happened anyway. A representative of the Asian Development Bank admitted that their first reaction to the CDM was to go through their portfolio to find already existing projects that they could gain additional money for through carbon credits.[31] In addition, carbon credits are awarded for maintaining carbon sinks in soils and tree plantations. The science behind these kind of projects is very uncertain and it is simply impossible to give an accurate idea of how much carbon is saved, if any. These spurious carbon credits are being used to compensate for real emissions in the North – this is like paying your gas bill with Monopoly money. Since CDM credits are two to four times cheaper than buying emissions credits through the EU ETS,[32] EU companies have every incentive to buy these cheap credits even if they are not worth the paper they are written on. This makes the whole system a sham.

Creating perverse incentives

Where it is expensive for a company to make the necessary adaptations to meet its emissions quota, it is more likely to buy credits from abroad than invest. Short term, bare-minimum compliance is incentivised over investment in long term emissions reductions in the North.

Southern governments are rewarded for imposing lax environmental standards so that baseline emissions scenarios are high and projects can gain large amounts of carbon credits. South Africa's Department of Mines and Energy has admitted that it faced pressure from the private sector not to make renewable energy targets too stringent, for fear future CDM projects will not be able to prove they are better than what would have happened otherwise.[33]

Carbon trading favours cheaper ways of reducing emissions. The cheap and easy changes are not necessarily the ones that are most important to make. Some emissions reduction measures can encourage a shift towards greater reductions in future. Other measures involve one off changes that do not lead to wider change or have additional social or environmental benefits. For example, a certain amount of emissions might be saved by improving the efficiency of a factory as a one off measure. The same amount of savings might be made by changing behaviour patterns to reduce demand for energy that would lead to vastly multiplied cuts in future. The type of emissions reductions has knock-on effects which can be at least as important as the overall level.[34]

Awarding valuable property rights for free to the worst polluters

The carbon trading system replaces the 'polluter pays' principle with the 'polluter buys its way out' or even the 'polluter profits' principle.

In Europe, rights to pollute have been given out for free to the 11,428 worst polluting installations.[35] Companies have had massive influence in the negotiating process, lobbying for a low overall cap and for high emissions quotas for their company, meaning they can stay below their cap easily and sell surplus permits at a profit to emitters in industries with less influence. In the UK, companies such as BP and Shell have been granted millions of pounds in windfall profits while the NHS is forced to pay tens of thousands of pounds to buy extra allowances.[36] In 2005, such 'exceptions' meant the total number of allowances awarded by the EU exceeded the cap by 10%. This translated to €1.8bn (£1.4bn) of free money for destructive corporations.[37]

This new system of private property rights is a major concern in and of itself. Governments are giving away rights to what should be seen as a public good. This is privatisation. History shows that when property rights are created they are more easily concentrated in the hands of the most powerful actors in society.[38] In this case heavy polluters are being awarded the right to pollute. This will entrench their expectations of further privileges when new emissions targets are negotiated. As the cap gets ratcheted down, these companies stand in a strong position for lobbying to retain their high allowances, leaving other sectors such as domestic heating, public sector bodies and the transport sector to bear the burden of higher emissions cuts. This also translates to the global level as Northern countries will fight to retain their high share of rights to emit.

Perpetuating reliance on fossil fuels

Carbon trading is aimed at making the dominant fossil-fuel-based model work more efficiently, not in pushing for radical change towards a lower carbon model. As Larry Lohmann points out: 'While trading schemes can in theory save participating private firms money in reducing emissions of specific substances, to a particular degree, over particular time periods, and within a particular larger technological system, the same schemes are unlikely to be the best choice if the objective is to save money for society or industry as a whole, or attain a more general environmental improvement, or make more drastic reductions with long-term goals in mind, or bring about a change in a larger technological system.'[39]
Carbon trading does little more than squander ingenuity and resources on making small refinements to technologies that need phasing out. This is also true of the CDM: 72% of CDM projects have been end-of-pipe add-ons at high-

emitting factories. Far from rewarding renewable energy projects with strong development benefits, the CDM in fact rewards 'well-financed, high-polluting operations capable of hiring professional validators of counter-factual scenarios'.[40]

As CDM expert Ben Pearson points out: 'A mechanism designed to promote climate protection...should be reducing the number of coal and oil projects, not providing them with a new revenue stream and diverting financing from renewable projects.'[41]

Unacceptable negative impacts on communities and the environment

Since heavy polluting sites can buy their way out of emissions cuts, carbon trading does nothing to improve the lives of local communities suffering as a result of other pollution associated with heavy industry and fossil fuel based power generation. These polluters are generally concentrated in poorer areas.

In addition, the cheap credits that allow companies to continue polluting poor communities in the North are all too often produced at the expense of communities in the South. CDM projects have been associated with land grabs, water and soil depletion, human rights abuses and pollution in communities. The CDM also fails to reward the many smaller-scale projects that offer real benefits for communities since they yield too few credits to be worth the cost of the bureaucracy.

Conclusion

Carbon trading has not pushed corporations into investing in the long term future of the planet. Instead, it has taught them how to lobby for more emissions permits, find ways of passing on costs to customers, locate cheap carbon credits abroad, present a green face to the public, keep fossil fuels such as gas as an option, and make marginal efficiency improvements.[42] Responding to and finding ways to profit from carbon prices is one thing; taking practical long-term action on climate change quite another.

Why corporations do not have the answer to climate change

Corporations may fool the market, but they can't fool the climate.

While it may be true that some corporations are beginning to find their own self-interest aligned with tackling climate change and are seeking to profit from the solution to the crisis, and while some individuals within corporations undoubtedly are genuinely concerned, the basic approach remains a problem.

Profit-maximising self-interest, even for a company trying to profit from solutions to climate change, is still a perverse incentive. Corporations have a history of making false claims in order to increase profits on every issue; from efficacy of pharmaceuticals, to levels of oil reserves, to the impact of their operations on communities in the global South. Publicity about technological solutions to climate change is no exception – hype wins over investors and sells products (for a while).

For example, in its 2001 sustainability report, Shell predicted that, in the medium term, drivers would be able to travel 400 km on two litre canisters of hydrogen bought from vending machines at garage forecourts.[43] This prediction bore no relation to developments in hydrogen technology (see below). The aviation industry is still using out-dated figures to proclaim that its impact on the climate is much lower than the public perceives it to be.[44] Emissions from agrofuel production are turning out to be even worse than from fossil fuel use, but corporations developing these fuels have turned a blind eye to the impacts in the pursuit of profit, while wrapping themselves in claims of being 'green'.

Such claims can delay investment in more useful technologies or cover up continuing emissions while the corporations profit. Carbon accountants might be convinced, emissions reductions credits may be sold, but as far as the climate is concerned the emissions are still there and climate change is no closer to being dealt with.

The Cleantech Boom

> *'When industry giants such as GE, Toyota, and Sharp and investment firms such as Goldman Sachs are making multi-billion dollar investments in clean technology, the message is clear. Developing clean technologies is no longer a social issue championed by environmentalists; it's a money-making enterprise moving solidly into the business mainstream.'*[45]

How the world of corporate-sponsored techno-fixes currently works can easily be seen from the markets. 'Cleantech' is set to be the new Dot Com, the new Biotech, the new technological wave that will change the world and make major profits for those companies that are able to ride the wave. Shaking off its environmentalist image in order to appeal to the business mainstream, 'cleantech' refers to 'any product, service, or process that delivers value using limited or zero non-renewable resources and/or creates significantly less waste than conventional offerings.'[46] It covers four main sectors: energy, transport, materials and water. In 2006, the clean energy sector alone was valued at $55.4bn (£27.8bn) by researchers at Clean Edge.[47] Researchers at Lux Research, experts in emerging technologies, have produced the first major analysts report into the sector and count the number of cleantech start-ups at 1,500 worldwide.[48]

Technology booms happen when a whole number of factors come together. Cleantech analysts describe a 'perfect storm' of factors which have contributed to its growth:

Climate change and peak oil
Rising oil prices, government demand for energy security and the potential costs of emitting greenhouse gases due to the carbon market have created incentives for business to invest in cleantech.

Emerging markets
Economic growth in China, India and many other countries in Asia, Africa and South America means investing in cleaner technologies, not least because there are not enough fossil fuels to go round. Millions of people are moving from rural areas to cities and economic growth means more people want access to electricity and other services, In 2006, China committed to investing up to $180bn (£90bn) over 15 years in clean energy.[49] Companies are chomping at the bit to maximise the opportunity to profit from serving these 'emerging markets'.

Technological convergence
New developments in nanotechnology, biotech and information technologies mean that there could be massive opportunities for the creation of new, more efficient and cheaper materials, processes and products.

Competition
Cities, regions and countries are beginning to compete to be the global hub for manufacturing and developing new clean technologies by offering subsidies and tax incentives.

Capital
Many of the investors that lead the growth of the high tech and internet revolutions have been investing in cleantech on the past few years. Vinod Khosla of Sun Microsystems is a key cleantech investor. Major companies such as Goldman Sachs, BP, General Electric have invested billions in the sector.

Consumer concerns
Consumer demand for clean products and services is increasing and helping to drive the market for specific consumer products.

Effects of the boom

Cleantech is a world away from niche alternative companies working to a social mission of improving the environment. This is hard-nosed business investment with a few environmentalist concerns tacked on for added PR. The valuation of cleantech companies is based largely on intellectual property and this is key to gaining investment from venture capitalists. The race is on to own the key technologies that future society will be utterly dependent on and to reap the royalties.

With this high-tech approach, performance is key. In the absence of encouraging legislation such as tax breaks or phased bans on damaging products, companies cannot market new products on environmental criteria alone, but have to match the performance of conventional products. So energy efficiency savings go into improving performance rather than saving energy, resulting in no overall reduction in energy use or increase in sustainability. This can be seen in car development. The Model T Ford had a fuel efficiency of 25 mpg. The average in the USA today is 21mpg. Efficiency gains have been used to increase speed, carry more weight and power more gadgets.[50]

For many, the cleantech 'boom' is a sign of optimism about the prospects for climate mitigation. The market is waking up, innovations are opening up new possibilities. But already analysts are warning that cleantech may be drowning under its own hype. Lux Research points to deals which are 'eerily reminiscent of the dot com boom' with companies being severely overvalued because more capital may be going into the sector than there are good enterprises to take it up.[51] This makes the market susceptible to boom and bust dynamics. However, at the time of writing the global credit crunch and impending recession in the USA means much of this money is drying up, which may slow the boom and rebalance the amount of capital available.

Of the thousands of start-ups only a small percentage are expected to make it big. For some of those developing the technologies it does not matter whether the technology works or not as long as they can attract funding, make money and sell up before the crash. This can be seen by looking at other sectors: for example, in the biotech boom of the 1990s, GM crops companies were promising food plants that produce drugs, or have enhanced nutritional content. Where are they now? In most cases, the companies have been bought up, merged and de-merged and the flagship projects that grabbed headlines ten years ago have disappeared from sight.

In the reckoning of the cleantech boom, those technologies that fail to make it will be the ones which are not easy to scale up and those for which companies fail to bring costs down to a marketable level. There are many reasons why a technology may fail this market test. The effectiveness and sustainability of the technology are only two of many factors, and as far as the market is concerned, not very important ones at that.

So the big companies and investors are jostling for position and playing a waiting game – looking to see which technology or exciting start-up will live up to its hype and waiting for the carbon price to reach a level that makes these technologies viable. Meanwhile solutions to climate change are hanging in the balance. Changes we should have started years ago are being delayed, useful technologies are witheld until their deployment is more profitable, emissions increase and the chance of avoiding dangerous climate change dwindles.

CHAPTER IV: THE TECHNOLOGIES - ENERGY PROVISION

Hydrogen

Hydrogen; the smallest and lightest element on the periodic table, and most abundant element in the universe. As humanity basks in the radiance of the vast hydrogen reactor at the centre of the solar system, hydrogen has a poetic rightness as an alternative fuel. It burns cleanly and the only exhaust gas is water vapour. All climate and energy security problems solved at a stroke.

Sadly, it's not that easy. Hydrogen is a carrier of energy not a source in its own right – a primary energy source – coal, gas or electricity generated from other sources – is required to produce it. Using hydrogen as a vehicle fuel (the main application being considered) would be colossally expensive to introduce, doing so would probably mean a commitment to long-term fossil fuel consumption and, most importantly, producing the hydrogen and compressing or liquefying it to use as a vehicle fuel could have a worse impact on the climate than using petrol.

How it works

Hydrogen can be produced from fossil fuels, including coal and gas, through a process of thermal decomposition. It can also be produced from water through electrolysis.

In vehicles, hydrogen gas can be burnt directly in an internal combustion engine, or used in a reaction with oxygen from air and a catalyst in a fuel cell to produce electricity to power a motor, with heat as a by-product. Very large cells used to power buildings can also use the heat for space heating or water heating, making them more efficient overall than smaller cells.

Hydrogen Current Position

Status of technology

Experimental. Not currently in commercial use as a transport fuel. There are key technical barriers to be overcome before hydrogen could be considered as a developed technology. These are in hydrogen production, transport, and storage technologies, as well as fuel cell technologies for use in vehicles. There are also significant safety concerns around the widespread use of hydrogen technology.

Support being received

The hype about hydrogen has died down with many governmental schemes introduced in the early 2000s now less active and much focus on future vehicle fuels switching to agrofuels.
In the USA, President George W. Bush announced '$1.2bn (£0.6bn) in research funding so that America can lead the world in developing clean, hydrogen-powered automobiles.'[52] in 2003. The US Department of Energy Hydrogen Programme supports research and development into hydrogen.[53] Their timeframe for adoption of the technology recognises that 'the hydrogen economy will take several decades to achieve'.[54]

The EU has provided €470m (£372m) of funding over the next six years for an industry-led programme of research, technology development and demonstration activities.[55]

Major players

Shell Hydrogen, Clean Urban Transport for Europe bus project. Most major car manufacturers produced hydrogen prototypes including: BMW, Honda, Toyota, Renault, GM, Ford.

Hydrogen Issues

Producing hydrogen usually requires fossil fuels

There are no naturally occurring deposits of pure hydrogen to tap into, unlike coal or oil. Hydrogen only comes bonded to other elements; it takes energy to separate it. Like a battery, it needs an energy source to produce it – gas, coal or electricity.

Hydrogen is not a new product at all. There is over a century's experience of industrial production (for nitrate fertilisers and oil refining). Much of what could be done to economise has been done. It is still very expensive.

Manufacturing hydrogen from natural gas is currently the cheapest method and the one with the longest track record. But gas is rapidly being depleted and a switch to using it for making large amounts of hydrogen will make gas run out even sooner.

More importantly, manufacturing hydrogen from gas has a major climate impact. In producing the hydrogen, the gas is split to form hydrogen and carbon dioxide. This 'clean' fuel only gives off water vapour from a car exhaust, but the carbon dioxide has already been emitted at the factory. This climate friendly renewable fuel is actually a carbon-emitting fossil fuel.

As gas is the cheapest way of making hydrogen, it's what manufacturers predominantly use and will continue to use. Manufacturing hydrogen from natural gas emits 9.1 kg carbon dioxide per kilogram of hydrogen.[56] Its climate impact is at least as bad as petrol,[57] even before taking into account the substantial extra emissions from liquefying or compressing it, then transporting it.

Hydrogen production using natural gas would only be low carbon if the carbon dioxide is captured and stored (see section on carbon capture and storage). The emissions from transporting the gas cannot be captured and these are increasing. As the European fields reach the end of their life, gas is being shipped in liquid form from the Middle East. To be liquefied, gas must be cooled to –163°C,[58] with all the energy consumption that implies, so the emissions from using gas effectively increase.

Hydrogen can be produced without using fossil fuels by electrolysis of water. Put simply, an electrical charge breaks the bond between the hydrogen and oxygen in H2O. This requires a great deal of energy, much more than producing hydrogen from coal or gas, and is also more expensive. This method is potentially also more carbon intensive, depending on the source of electricity. Powering BMW's new hydrogen car with electrolysis hydrogen using electricity from the UK grid would create around four times the emissions of its petrol equivalent.[59]

The only genuinely carbon-free hydrogen would use renewable electricity to power electrolysis of water. Creating enough hydrogen by this method to supply a significant proportion of vehicle fuel use would massively increase the overall demand for electricity.

Using electricity to produce hydrogen is much less efficient than using electricity directly. Using renewable electricity to produce hydrogen would save 220kg of carbon dioxide emissions per megawatt-hour of final energy output compared to electricity generated from fossil fuels.[60] Using renewable electricity directly as electricity saves 370kg of carbon dioxide if it replaces gas-fired generation, and 890kg if it replaces coal-fired generation.

Replacing the UK's vehicle fuels with electrolysis hydrogen would take more than the country's present electricity consumption.[61] Is it possible to double electricity generation while doing away with fossil fuel use? Or is electrolysis hydrogen as a vehicle fuel a non-starter?

When the Bush administration used the 2003 State of the Union address to announce the kickstart of the hydrogen economy for vehicles, they neglected to mention any of the emissions that come with making it. Under the National Hydrogen Energy Roadmap plan, fossil fuels would be the source of the vast majority of hydrogen, but 10% would come from electrolysis of water, powered by dedicated nuclear power plants.[62]

Timeframe

Dr Joseph Romm ran the US Department of Energy's Office of Energy Efficiency and Renewable Energy from 1993-98, during period when the 'hydrogen economy' became big news. Yet, although Romm is a confessed believer in the possibility of clean hydrogen, he is firm in his belief that in 2030 less than 5% of vehicles will be powered by it.[63] Even if the technologies were available for hydrogen to be able to reduce greenhouse gas emissions, this timescale is far too long for investment in hydrogen to be a practical way out of the current emergency.

Energy required to compress or liquefy makes it entirely impractical as a vehicle fuel

At room temperature and pressure, hydrogen has one three-thousandth of the energy of petrol.[64] To be practical as a vehicle fuel, hydrogen needs to be either compressed or liquefied. To be liquefied, hydrogen needs to be cooled to -253°C.[65] The energy used to do this is equivalent to 30-40% of the energy the hydrogen contains.[66]

It takes 12.5-15 kilowatt-hours (kWh) of electricity to liquefy 1kg of hydrogen.[67] With the UK's emissions from generating electricity, that's 6kg-7.2kg of carbon dioxide emissions.[68] A gallon of petrol – which contains around the same amount of energy as 1kg of hydrogen[69] – releases around 8.8kg carbon dioxide.[70] In other words, hydrogen causes 68-82% of the emissions of burning petrol just for the liquefaction process.

Hydrogen can also be left at room temperature as a gas but compressed. This is Honda's choice for their FCX model. Compression takes less energy than liquefaction, but compressed hydrogen contains less energy than liquefied hydrogen. By volume, at the sort of pressure likely to be used in cars, hydrogen gas contains around one-tenth of the energy of petrol.[71] This means the car has a fraction of the driving range of a petrol or diesel car. It won't just need refuelling more often, it also means that more tankers, pipelines and other infrastructure would be required. A study examining the possibility of trucks carrying compressed hydrogen in tubes showed that for every 200 miles the truck drives, it uses energy equivalent to 20% of the fuel it delivers.[72]

As a liquid, hydrogen needs to be maintained at -253° centigrade constantly until the point of use. This takes a further significant quantity of energy. By now, not only are the emissions about the same as burning petrol, but around half the energy has effectively been lost.

There are certainly good cases for having a few very inefficient devices – it's far handier to power a camera off a battery than a small diesel motor. But using grossly inefficient technology for such a huge energy consumer as the world's vehicle fleet would be wildly impractical.

As electric vehicle advocate Alec Brooks put it, '...fuel cell vehicles are energy pigs. Fuel cell vehicles that operate on hydrogen made with electrolysis consume four times as much electricity per mile as similarly-sized battery electric vehicles.'[73]

BMW's hydrogen vehicle, the H7, uses hydrogen not to power a fuel cell, but directly in an internal combustion engine. Being even less efficient than a fuel cell, it means it has a smaller driving range, a mere 125 miles on eight kilos of hydrogen.[74] The fuel is kept in liquid form in an insulated tank. The insulation cannot prevent the fuel warming, only slow it down. The fuel gradually reverts to gas as it warms and pressure in the tank increases. This is not a safety issue as, like a petrol car, a valve allows it to escape. It is, however, an economic issue. The H7 is preprogrammed to start jettisoning fuel after less than a day. A full tank takes less than a fortnight to empty itself.[75]

Joseph Romm, despite being a hydrogen advocate, is incredulous at the BMW design, commenting, 'BMW has managed to develop the least efficient conceivable vehicle that you could invent'.[76]

This inefficiency means that it is more expensive for the owner, worse for the climate and not a practical solution. As David Talbot put it in Technology Review, '...a car like the Hydrogen 7 would probably produce far more carbon dioxide emissions than gasoline-powered cars available today. And changing this calculation would take multiple breakthroughs - which study after study has predicted will take decades, if they arrive at all. In fact, the Hydrogen 7 and its hydrogen-fuel-cell cousins are, in many ways, simply flashy distractions produced by automakers who should be taking stronger immediate action to reduce the greenhouse-gas emissions of their cars.'[77]

Cost of infrastructure

With current technology, installing the infrastructure to supply just 40% of the light-duty vehicles in the USA alone has been estimated to cost over $500bn (£250bn).[78]

Don Huberts, CEO of Shell Hydrogen, confirms this estimate: 'The initial investment has been estimated by Shell at around $20bn (£10bn) for the U.S. alone, to supply 2% of the cars with hydrogen by 2020 and to make hydrogen available at 25% of the existing gasoline retail stations. In the subsequent decades, further build-up of the hydrogen infrastructure will require hundreds of billions of US dollars.'[79] Much greater emissions reductions could be achieved elsewhere for these amounts of investment.

The cost to the consumer

At present, hybrid petrol/electric cars are doing well commercially because they are more economical. It's easy to sell someone the environmentally friendly option when it's going to be cheaper for them. No driver, irrespective of their environmental feelings, will buy a car if they can't afford it.

In December 2002, Yozo Kami, Honda's engineer in charge of hydrogen fuel cells, said it would take at least ten years to get the price of a hydrogen car down to $100,000 (£50,000). Honda's is one of the cheapest prototypes.[80]

Fuel cells of the type used in cars (proton exchange membrane cells) have a short lifespan. The industry is aiming at around 4,000 hours of use, which might equate to ten years of driving. As it stands, a good prototype can only manage about 2,000 hours.[81] Buying a car that costs £50,000 and will be useless in five years isn't going to appeal to anyone.

And even once the industry were scaled up, the costs of infrastructure for hydrogen as a vehicle fuel are still likely to be at least twice high as for petrol.[82] In 2003, Shell Hydrogen's CEO Don Huberts bluntly conceded, 'at the end of the day, hydrogen and other alternative fuels will be three to four times as expensive as oil based products, and if no-one wants to pay for that, we can't make those fuels.'[83] To some extent this effect may be offset by rising oil prices, but, again, if this is the best possible result, would the investment be more useful elsewhere?

Safety

Being a small molecule and very light, hydrogen is particularly leak prone. It is also odourless. Natural gas can have varying or no odour, so an odorising agent is added to it. But odoriser cannot be added to hydrogen as not only might it damage the fuel system technology (especially in sensitive fuel cells) but it wouldn't actually work - the hydrogen would be substantially lighter and would separate from its smell.

Worse still, hydrogen is not only leaky, invisible and odourless, but it burns invisibly too. A raging fire could be undetectable until you stepped into it and went up in flames. NASA developed a way of detecting leaks in its hydrogen tanks by getting someone to walk round pushing a broom in front of them to see if the bristles caught fire.[84]

It is a very dangerous substance to be handling in large quantity, in populated areas, at thousands of forecourts with untrained members of the public.

Hydrogen is flammable over a wide range of concentrations and has ignition energy twenty times smaller than natural gas or petrol. 'Operation of electronic devices (cell phones) can cause ignition',[85] and 'common static (sliding over a car seat) is about ten times what is needed to ignite hydrogen'.[86] Electrical storms several miles away can generate enough static for ignition.[87]

But surely, with it being so volatile and also with it being a new technology needing public confidence, the motor manufacturers have taken extra care and got all this covered, right? Wrong. In May 2003 Toyota recalled all its prototype hydrogen vehicles after a leak was discovered in the tank of one. Not by engineers, but by the driver noticing a strange noise when refuelling.[88]

Safety is not just an issue for filling stations and vehicles. Tankers on the roads are also a problem. Because compressed hydrogen carries so little energy value, more tankers would be needed. A study in the USA suggests that fifteen times the number of tankers would be needed compared to supplying petrol, and one out of seven accidents involving trucks would involve a hydrogen truck. Every seventh truck-truck collision would occur between two hydrogen carriers.[89]

All the new infrastructure would need serious safety testing before rolling out nationally and globally. Even with a massive and unflinching political, industrial and financial push, the turnaround would take a couple of decades. Even if hydrogen offered a means of making major cuts in greenhouse gas emissions, there isn't that sort of time available.

Hydrogen is highly reactive

Hydrogen bonds easily with other substances. Its reactivity causes metals, including steel, to become brittle. Pipelines, tanker trucks and other storage and supply equipment would need to be made of higher grade materials and/or replaced more frequently. The infrastructure costs could be astronomical.

Higher strength materials are more susceptible to hydrogen embrittlement.[90] Lower strength materials are, of course, more susceptible to rupture. There are some possible ways around this. There are proposals to have filling stations producing their own hydrogen. This would remove the delivery fleet and pipeline issues. However, putting a hydrogen production plant in every filling station would be phenomenally expensive and, with production dotted at thousands of small sites scattered round the country, completely rule out any chance of carbon capture and storage if fossil fuels were being used (see section on carbon capture and storage). Also, the smaller scale plants at filling stations would be even less efficient than making hydrogen centrally, cancelling out much of the saving from avoiding transport.

Astonishingly, some manufacturers are looking at on-board hydrogen manufacturing from a variety of sources. As the manufacture of hydrogen takes place in the car itself, there is no possibility of carbon capture and storage. Perhaps the most preposterous suggestion is for the use of petrol as the source fuel. Rather than burning it directly, the car uses petrol to make hydrogen as fuel for its fuel cell. In 2004, Renault was reported to be working with a company called Nuvera Fuel Cells to develop this.[91] The energy consumption and greenhouse gas emissions of such a vehicle would be greater than for a hybrid petrol vehicle.[92]

Hydrogen may commit us to coal

Hydrogen requires large amounts of energy to produce, large amounts of energy to store and large amounts of energy to distribute. To increase the demand for natural gas by making it the primary fuel for hydrogen means the price of gas would go up, making coal even more viable for electricity generation.

Yet the gas supply is only temporary too. Investing hundreds of billions in infrastructure to make a global hydrogen economy means a commitment to many decades of hydrogen use. When gas supplies peak, predicted for the middle of this century, coal will become the cheapest way of making hydrogen.

Hydrogen has been spun beyond all relation to reality

Shell is the most visible company pushing the idea that Iceland will be 'the world's first hydrogen economy,' implying that Iceland is leading the way and other countries can follow. In April 2003, Shell opened a hydrogen filling station in Reykjavik, and proudly said that they had immediately signed up three of the city's buses.[93]

Iceland is a peculiar case, not only peculiar because it is sat on more renewable energy than it can use (a few huge hydroelectric plants and lots of geothermal energy); it is also little more than a city state. It has a population of 300,000 – the size of Bradford – and two-thirds of them live in one city. Those first three buses are 4% of the Reykjavik's entire fleet. So three or four filling stations cover most of the country. That simply cannot be scaled up to the UK, or anywhere else.

In 2001 Shell produced a report called *Energy Needs, Choices and Possibilities: Scenarios to 2050*. It describes several possible futures. One talks of 'developing a new "fuel in a box" for fuel cell vehicles'.[94] BMW's admittedly inefficient hydrogen combustion car the H7 uses over a litre of fuel to go 2km, yet in Shell's scenario 'a six-pack of fuel (12 litres) is sufficient for 400 km,' or over 16 times the efficiency of the H7. Even more curiously, it talks of this fuel in a box as being 'distributed like soft drinks through multiple distribution channels, even dispensing machines'.[95]

For Shell, all the safety problems of having hydrogen vending machines seemingly don't exist. Presumably they disappeared around the same time as this mysterious safe canister material was invented. Shell neglects to say whether these will be high-pressure gas canisters (remember, strong containers embrittle faster, weaker ones break easier), or ones holding liquid at -253°C (necessitating a huge electricity bill for the vending machine, or else a super-insulated

Scores: Hydrogen

Effectiveness at reducing greenhouse gas emissions: 1/10
Living up to the hype (science to spin ratio): 1:10
Democratic ownership and control: 1/10 – perpetuates dominance by big energy corporations
Social justice: 3/10 – no new side-effects, but perpetuates existing energy inequality
Sustainability: 2/10 – even if not reliant on fossil fuels, perpetuates negative impacts of vehicle use.
Scalability: 2/10 – massive infrastructure needed, even more expensive (therefore elitist) than current vehicle technologies

canister that leaks the fuel and, if it's been sat in some underused vending machine, may be empty by the time it's bought). One of Shell's current slogans is 'real energy solutions for the real world'.[96]

Shell appears to be deliberately ignoring the serious and currently insurmountable engineering and safety problems of hydrogen. Its scenarios could not possibly be in place by its projection of 2025. Has the company that invented greenwash been taken in by its own hype or is this just a decoy to distract consumers from the need to reduce fossil fuel use?

Conclusion

For hydrogen to be viable as a vehicle fuel it needs numerous technological breakthroughs in all major areas including production, distribution and storage. The fuel and vehicles, even with some optimistic assumptions about technological breakthroughs, will be more expensive than conventional vehicles to buy, more expensive to refuel, and will not last as long.

For hydrogen to reduce greenhouse gas emissions would require a glut of renewable electricity or universal carbon capture and storage within a decade. The likelihood of the infrastructure that would make it work being put in place within four or five decades is slim. So, the technical and economic issues mean it cannot be a climate change solution.

BMW's Hydrogen 7: emissions comparison

Producing the hydrogen:

- Electrolysis requires 39 kWh of electricity to produce 1 kilogram of hydrogen.[97]
- The BMW H7 has an 8kg hydrogen tank, giving a 200km driving range.[98]
- 39 kWh electricity to produce 1kg hydrogen x 8kg capacity of tank = 312 kWh electricity per tank
- 312 kWh electricity divided by 200 km driving range = 1.56 kWh electricity per km driven.
- UK grid electricity has carbon dioxide emissions of 480g/CO_2 kWh.[99]
- 480g/CO_2 per kWh x 1.56 kWh = 749g/CO_2 per km to make the required hydrogen gas from electrolysis.

Manufacturing hydrogen from natural gas emits 9.1kg CO_2 per kg hydrogen [100]
9,100g/CO_2 x 8kg H2 in the tank = 72,800g/CO_2 per tank
72,800 divided by 200km driving range = 364g/km to make the required hydrogen gas from natural gas.

Then it has to be liquefied:

- It takes 12.5-15 kWh of electricity to liquefy 1kg of hydrogen.[101]
- 12.5-15 kWh electricity x 480 g/CO_2 per kWh = 6-7.2kg/CO_2 per kg hydrogen liquefied.
- With a tank capacity of 8kg and a driving range of 200km the car travels 25km per kilogramme of hydrogen.
- 6–7.2kg/CO_2 for 25km or 240-288g/CO_2 per km for liquefaction

In total:

749g/CO_2 per km for hydrogen production + 240-288g/CO_2 per km for liquefaction =
989 - 1033g/CO_2 per km for liquefied hydrogen produced via electrolysis
or
364g/CO_2 per km for hydrogen production + 240-288g/CO_2 per km for liquefaction =
604 - 652g/CO_2 per km for liquefied hydrogen produced from natural gas

For comparison:

The petrol car the H7 is based on, the BMW 750, emits 271g/km.[102]
A Toyota Prius emits 104g/km, a Renault Megane emits 117g/km, a vicious gas guzzler like the Porsche Cayenne emits 310g/km.

Agrofuels

Introduction

Agrofuels have been trumpeted as the 'green' solution to cutting oil use for transport, but in fact threaten an environmental and social justice disaster.

Agrofuels rely on industrial scale agriculture, the growth of which has been dependent on deforestation and cheap fossil fuels. The dream of swapping fossil fuels with energy from plants fundamentally misunderstands what a fossil fuel is. It took four centuries worth of plant and animal growth to create the fossils we burn in a year.[103] The biosphere is simply incapable of creating enough energy to support this kind of demand in the long term. Providing that amount of energy would require minimum 50% increase in human use of biosphere productivity, before factoring in any loss of energy for conversion to fuels.[104]

Agrofuels is a term coined by critics of the way that the term 'biofuels' is used to put a green spin on what is essentially a new growth area for agribusiness giants such as Cargill and Archer Daniel Midland.

How does the technology work?

Agrofuels are liquid fuels produced from agricultural crops, primarily used for road transportation.

'First generation' agrofuels include bioethanol, which is used as a replacement for petrol, produced by fermenting and distilling the sugars in feedstocks such as maize or sugarcane, and biodiesel created from vegetable oils, mainly soya, rapeseed oil, or palm oil.

'Second generation' agrofuels mainly use one of two processes. The first converts cellulose from feedstocks such as wood, switchgrass or agricultural waste into ethanol. The second uses gasification and the Fischer-Tropsch process to convert various types of biomass to biodiesel. The carbon component of the biomass is gasified in steam, producing hydrogen and carbon monoxide ($C + H2O \rightarrow H2 + CO$) which the Fischer-Tropsch process converts to hydrocarbons and water.

Another branch of second generation (or 'third generation') agrofuel research involves growing large quantities of algae, some species of which contain relatively large proportions of fats and oils, and converting them to biodiesel or biobutanol.

Agrofuels Current Position

Status of technology

First generation agrofuels are on the market and currently expanding rapidly. Second-generation agrofuels are not yet on the market, with most still at the experimental or speculative stage. For the most part, agrofuels are not currently price-competitive with fossil fuels except when subsidised. Second generation agrofuels require breakthroughs in genetic engineering of plants, microbes and enzymes to increase the efficiency of processes to break down feedstocks and increase yields per hectare of land. Algae production appears to work technically but faces huge obstacles to commercial viability.

Proportion of current use

Less than 1% of global vehicle fuel use.

Support being received

The global market for agrofuels has boomed in a short time, helped by government subsidies and targets to replace a proportion of fossil-based transport fuel with agrofuels.

The UK government has introduced a 'Renewable Transport Fuel Obligation' requiring 5% of road transport fuel to be replaced with agrofuels.[105] On 21st February 2008, Transport Minister Ruth Kelly announced that this policy was under review, stating that; 'we are not prepared to go beyond current UK target levels for biofuels until we are satisfied it can be done sustainably.'[106]

The EU's current target is for 5.75% of road transport fuel to be from agrofuels by 2010,[107] and in March 2007 a new target of 10% by 2020 was announced.[108] An unpublished paper from the European Commission's own Joint Research Centre has criticised this new target saying that 'the uncertainty is too great to say whether the EU 10% biofuels target will save greenhouse gas or not.'[109] In January 2008, the EU's Environment Commissioner Stavros Dimas admitted that the negative impacts of hitting the target on communities and the environment had not been properly thought through, though the target has not been dropped.[110]

The US Energy Policy Act 2005 mandated the substitution of 7.5 billion gallons of gasoline per year with agrofuels. In his 2007 State of the Union address, President Bush announced a mandatory fuels standard to require 35 billion gallons of agrofuels in 2017. The production of maize based ethanol in the USA is projected to reach 5.9 billion gallons in 2006/07 and 9.7 billion gallons in 2010/11.[111] The primary motivation for this is lessening reliance on Middle Eastern oil, not solving climate change.

Jean Ziegler, the United Nations Special Rapporteur on the right to food, has called for a five year moratorium on agrofuel production.[112]

Major players

Agrofuel companies (UK): Refining and distribution companies currently operating or with plants under construction:
Argent Energy, Associated British Foods, D1 Oils, Earl's Nook Ltd. (formerly Biofuels Corporation), Ensus, Greenergy, Green Spirit Fuels, Ineos, Abengoa

Oil companies - Investment, refining and distribution:
BP, Shell, Petrobras, Conoco Philips

Agribusiness - supplying feedstock :
Cargill, Archer Daniels Midland, British Sugar, Pioneer

Second generation agrofuels - Biotech and synthetic biology companies:
DuPont, BASF, Syngenta, Monsanto, Novozymes, Synthetic Genomics, Amyris Synthetic Biology, LS9, Macoma

Investors:
Khosla Ventures, Virgin Fuels

Agrofuels Issues

Though agrofuels currently supply only a very small percentage of world transport fuel use, their impacts are already significant and the full implications of replacing fossil fuels with agrofuels are beginning to be seen. Rainforests are being destroyed in Brazil, Malaysia and Indonesia as agrofuel plantations expand, destroying carbon stores which are vital to regulation of ecosystems. Competition for land between food crops and fuel crops is already contributing to rising global food prices, with food crises in 37 countries. Subsistence farmers are losing their land and rising prices are causing riots around the Majority World. Evidence has emerged that many agrofuels ultimately increase rather than reduce greenhouse gas emissions.[113]

Climate impact

Agrofuels have been promoted as 'carbon neutral' because the carbon released when the fuel is burnt is largely equivalent to that absorbed from the atmosphere as the plant grows. However, agrofuels still have a significant climate impact which is often greater than the impact of burning fossil fuels.

New studies published in Science have shown that converting forest or grassland to grow agrofuels, or displacing current agricultural activity onto new land, leads to greenhouse gas (GHG) emissions that, in the cases investigated, are significantly worse than the fossil fuels they were supposed to replace.[114] These results have been sufficiently compelling to prompt the UK government to begin a review of its policy on agrofuels.[115]

Agrofuels' impacts on the climate include energy used to refine and transport agrofuel and feedstock, fuel for agricultural machinery, and nitrous oxide emissions from fertilisers used on crops. In addition, even more serious impacts are created by the expansion of agricultural land to grow agrofuel crops.

These impacts are caused by deforestation, peat drainage, soil degradation and forest fires. Deforestation releases the large amounts of carbon stored in forest vegetation – the IPCC estimates that a fifth of current greenhouse gas emissions result from deforestation. Peat drainage allows waterlogged peat soils to dry out and decompose, releasing massive quantities of stored carbon. A report by Wetlands International shows that every tonne of palm oil produced on peatland results in up to 33 tonnes of carbon dioxide emissions, or 10 times as much as petroleum.[116] Forest fires (often set deliberately to clear land) release carbon from vegetation and soils very quickly; in 1997/98, fires partly set by palm oil plantation owners spread across millions of hectares of degraded peatlands and forest in Indonesia,[117] releasing greenhouse gas emissions equivalent to up to 40% of annual global emissions from fossil fuels for the 1990s.[118]

The feedstock with the highest greenhouse gas emissions is palm oil, as this is most likely to be associated with deforestation and peatland conversion. Malaysia and Indonesia are roughly equal as the world's largest producers of palm oil.[119] Indonesia has the highest deforestation rate of any major forested country and oil palm plantation expansion has been named as the main driver of this,[120] with 6.4 million hectares of plantations already and 20 million hectares expansion planned over the next 20 years.[121]

Soils and plant biomass are the two largest biologically active stores of terrestrial carbon, together containing approximately 2.7 times more carbon than the atmosphere.[122] Where converting wild ecosystems to cropland releases carbon dioxide, this is known as the 'carbon debt' of land conversion. If an agrofuel crop has less total greenhouse gas emissions than the fossil fuels it displaces then this 'carbon debt' may be paid back over time. Until that point agrofuels from converted lands have greater greenhouse gas emissions than the fossil fuels they displace.[123] A recent report for the UK Treasury found that when tropical forest is cut down, 'Two to nines times more CO2 is released than would be saved by using an equivalent area of land to grow biofuels for 30 years. Put another way, to pay back the initial release of CO2 [the carbon debt] from clearing the forest would take 60 to 270 years of growing biofuels (using current technologies).'[124]

It's not just palm oil. One study shows that US corn-based ethanol, instead of producing a 20% greenhouse gas saving, nearly doubles emissions over 30 years and increases greenhouse gases for 167 years. Agrofuels from switchgrass, if grown on US corn lands, increase emissions by 50%.[125]

Competition for land - food vs fuel vs forest

Replacing even small quantities of oil with agrofuels requires a colossal quantity of feedstock and huge areas of growing land. Calculations by Greenpeace based on figures from the US Department of Agriculture and the International Energy Agency show that replacing just 10% of world diesel demand for road transport would require over 76% of total current global soya, palm and rapeseed oil production.[126]

Agrofuels do not have to be grown directly on virgin rainforest to be unsustainable. If agrofuel production uses land that is already under cultivation it displaces the agricultural activity that was already going on to somewhere else, and eventually, somewhere, a habitat is lost as more land is brought under cultivation. For example, the UN Food and Agriculture Organisation reports that EU palm oil imports have doubled since 2000, mostly to replace domestic rapeseed oil diverted from food to fuel use. This is driving further expansion into forest and peatland in Indonesia and Malaysia.[127]

Continuing on the current development track for agrofuels is likely to cause famine. Humanity faces a triple whammy in competition for land as populations rise, crops are diverted for agrofuels, and land and yields are lost to soil degradation and the effects of climate change - including sea level rise, drought and changing weather patterns. People in the Majority World are already suffering most from the impacts of climate change, and are having increasing difficulty buying food as agrofuel expansion causes food prices to sky rocket. As the UN's Food and Agriculture Organisation has pointed out, food will lose out to fuel in competition for land because the potential market for energy is so huge.[128] Those who own cars are richer than those who do not.

At the time of writing, the World Food Programme is warning that it will have to cut food aid due to high food prices unless governments donate an additional half-billion dollars in 2008.[129] Josette Sheeran of the World Food Programme said, 'This is the new face of hunger ...There is food on shelves but people are priced out of the market. There is vulnerability in urban areas we have not seen before. There are food riots in countries where we have not seen them before.' They attribute 30% of this change to agrofuel expansion.

In defending food sovereignty – the ability of people to have control over their food supply – localisation and control of production are key. Yet the drive for agrofuels runs totally counter to this, concentrating land ownership in the hands of large producers, increasing dependency on unstable global commodity markets and removing people from their land resulting in the loss of knowledge of local subsistence farming methods.

For agrofuel feedstocks to be grown, a choice must be made between destroying ecosystems or taking over land currently used for growing food. Neither is a sustainable proposition.

Biodiversity

In March 2008, the Royal Society for the Protection of Birds (RSPB) launched an advertising campaign on agrofuels with the dramatic strapline 'Don't put wildlife in your tank'. Land use change driven by agrofuel expansion is of critical concern not only due to the loss of carbon stores but also because of the loss of habitats and associated biodiversity. Tropical rainforests are areas of high biodiversity, yet as yields for typical agrofuel feedstocks are highest in the tropics, particularly for palm oil, soya and sugar cane, these important habitats are under threat.

Other key habitats under threat due to agricultural expansion include the Cerrado in Brazil,[130] the worlds most biodiverse savannah, and the Pantanal in Latin America,[131] the world's largest tropical wetland area. In Europe, the policy of 'set aside', which has led to subsidies for farmers returning a proportion of their land to wildlife, looks to be abolished due to rising grain prices and pressure for agricultural expansion.[132]

Monoculture

Since agrofuel refineries require huge volumes of uniform feedstocks at the lowest possible price, agrofuel technology favours large scale, intensive and unsustainable farming methods. Monoculture farming has been described as creating 'green deserts'; vast areas of land with no biodiversity where one single crop is grown.[133]

Monoculture is energy intensive, heavily reliant on artificial inputs of fertiliser, herbicide and pesticide, pollutes land, rivers and seas, causes large scale habitat destruction and a loss of biodiversity, depletes soils of nutrients, depletes water supplies in areas where irrigation is necessary, concentrates land ownership and causes rural poverty and migration to cities.[134] Throughout the global south, civil society groups are resisting the expansion of monoculture plantations in favour of sustainable farming practices and food sovereignty. Agrofuel expansion is heading in exactly the wrong direction.

Second generation agrofuels do not solve the problem

For many agrofuels advocates, second generation agrofuels, which convert cellulose into ethanol or produce fuel by gasifying biomass, are an answer to many of the sustainability issues associated with first generation fuels. However, second generation agrofuels face many technical barriers, are not proven to reduce greenhouse gas emissions, and do not solve the issues of food security or land competition. They also raise new concerns around the use of genetic engineering, synthetic biology and increased corporate concentration. Just as in the past oil companies bought solar companies,[135] promising future renewables expansion to distract people from the destructive impacts of oil, the promise of more sustainable second generation technology in the future is serving as an excuse for the unsustainable expansion of first generation agrofuels today.

Technical barriers

To produce fuels from non-food crops, enzymes or microbes must be used to break down cellulose. Cellulose is a difficult substance to work with and no effective system exists to easily convert it into a fuel. Once cellulose is broken down by enzymes it forms a mixture which is very dilute and must be distilled. The methods for pre-treatment and distillation are very energy intensive. Cellulosic ethanol is currently 'more costly and less efficient' than conventional corn ethanol.[136] Commercial cellulosic ethanol production is currently very limited but growing. Canada's Iogen Corporation, in partnership with Shell, runs the first commercial plant which opened in 2004.[137]

Any improvement on this requires a technological breakthrough, with researchers looking into genetically engineering enzymes and microbes, and even creating entirely new microbes from scratch (termed synthetic biology), to break down cellulose. Synthetic biology company LS9 is designing entirely new microbes that convert feedstocks into a range of 'designer' hydrocarbon-based agrofuels.[138] These technologies are a major focus of investment for both venture capitalists and governments. This is at the expense of investment into other emissions-reducing technologies. The US Department of Energy has diverted its entire budget for geothermal and advanced hydropower research into second generation agrofuels[139] and the EU is giving stronger policy support to agrofuels than any other type of non-fossil fuel energy.

Using 'marginal land', non-food crops and 'agricultural waste'

Second generation agrofuels are supposed to be more sustainable because they are less likely to use food crops as a feedstock (so should not increase food prices) and can make use of land that is not appropriate for cultivation of food. This is the theory. In practice, 'marginal land' is a concept used only by people who do not live on it. 'Marginal land' is currently under severe stress in many places.[140] Since such land is often used as pasture for livestock, second-generation feedstock production is still likely to displace agricultural activity, as well as damage ecosystems and biodiversity. Even if the feedstocks are not food crops, because of the volume of feedstock needed for refineries, second generation agrofuels will still rely on intensive monoculture farming methods and still increase competition for land which will drive up food prices.

Another proposed alternative is to use 'agricultural waste'. A joint report by the US Departments of Energy and Agriculture speaks of using 1.3 billion tonnes of dry biomass per year for 'renewable energy'. This would require most of the USA's agricultural wastes.[141] However, using 'waste' removes nutrients which in sustainable farming practices would be returned to the land. Removing nutrients and organic matter without replacing them 'mines' the topsoil. Organic matter returned to soil prevents erosion, improves soil structure and water retention, and gives the next crop its nutrition. Modern agriculture only addresses the nutritional component by adding fossil-fuel based - and greenhouse gas-emitting - fertilisers and because the soil is unhealthy from a lack of organic matter, copes with insects and disease with oil-based pesticides.[142] Using agricultural waste will also increase the costs and emissions associated with transporting feedstocks to refineries because sources will be widely dispersed.

Genetic modification (GM) and synthetic biology

Agrofuel development is a major new area for biotech companies.[143] Agrofuels provide a new market for existing GM crops including soya, maize and oilseed rape. The use of these technologies has been controversial for food crops and also, to a lesser extent, for animal feed crops, but the industry is hopeful that there will be less resistance to their use as a fuel. New varieties are being developed specifically for use as agrofuels: in some cases the emphasis is on increased yields, in others on designing crops that have properties that make them more suitable for using residues as animal feed.[144] Biotech companies are also seeking to engineer plants which have higher cellulose content; new enzymes and microbes to break down cellulose; and plants whose cellulose can be broken down more easily.[145]

The use of GM technology for agrofuel production is worrying not only because of the potential for the spread of genetic contamination, but also because existing GM crops rely on unsustainable industrial-scale monoculture farming practices including widespread crop spraying. Many existing crops are developed specifically to work with a proprietary herbicide (eg Monsanto's RoundUp Ready varieties). Use of herbicides is associated with increased skin and respiratory ailments, and cancer in workers and people living near the crop, and poses a significant threat to biodiversity. Despite the negative impacts, companies promote these crops as low carbon because they can be (though are not always) planted without tilling the soil – less tilling can (but does not always) reduce release of carbon dioxide from the soil.[146]

One particularly area of concern is the research going into genetically modified trees as a feedstock for second generation agrofuels. Poplar, eucalyptus and willow are all potential feedstocks and are being modified to grow faster and to have high cellulose and low lignin content (lignin strengthens the cell walls of plants making it harder to break down the cellulose).[147] All the problems with GM are particularly acute with trees as they have long lifespans; their pollen and seeds travel long distances, and many plants and animals depend on trees within an ecosystem.[148]

In addition to conventional genetic modification, agrofuels investment is the major source of funding for a new breed of bioscience companies developing entirely new life forms. Gene maverick Craig Venter's new company, Synthetic Genomics, is developing a new microbe with lucrative commercial applications. The synthetic 'minimal genome' microbe based on mycoplasma genitalium[149], dubbed 'Synthia' by genetics campaigners,[150] acts like computer hardware. The theory is that short strings of genes could be inserted into Synthia's simple genome to make the microbe act like a miniature factory, engineering its metabolism to convert specific substances into other substances which could have commercial applications.[151] Venter predicts that this could be 'the first billion or trillion dollar organism'. [152] Research into this area has been supported by the US Department of Energy's Genomes to Life programme. [153] The former head of this programme, Aristides Patrinos, is now president of Synthetic Genomics.[154]

The company hopes to put its new lifeform to work on the creation of 'second generation' agrofuels. Synthetic Genomics also hopes that the microbe will in future perform other functions including creating hydrogen, sequestering carbon dioxide and producing drugs.[155] Synthetic Genomics has recently gone into partnership with BP to study microbiology in oil fields to, amongst other things, improve methods for recovering oil.[156] Other companies working in this area include LS9, Mascoma and Amyris.

With all the government subsides and venture capital sloshing around for 'cleantech' and talk about the possibilities synthetic biology holds for climate mitigation and curing disease, some big issues are being left out. Synthetic biology (synbio) has been called 'genetic engineering on steroids'[157] and any concern people may have over GM applies equally to synbio. And just as biotech companies tried to sell the public the idea of GM on the basis that it would feed the world, synbio is also being sold on the issue of the day – climate change. Issues being ignored include the potential for the technology to be used to create bioweapons.[158] With governments desperate for promises of solutions that don't impact on levels of consumption, the time is ripe for controversial technologies to advance with little regulatory oversight.

Certification will not solve the problem

In response to the backlash against agrofuels, governments including the UK and EU are considering the introduction of sustainability criteria for agrofuels and applying certification to sources deemed sustainable. However, the proposed criteria will do nothing to increase the sustainability of agrofuels since they do not tackle the issue of agricultural expansion. An effective system would have to guarantee that feedstocks would not trigger significant emissions from land use change, for example by using a waste product or growing the feedstock on carbon-poor lands. This would exclude the majority of commercially available agrofuels.

Biofuels vs biomass

Above all, any biomass that can be produced sustainably would be more efficiently used generating heat and electricity than being converted into a liquid fuel by energy-intensive processes. One study of typical EU-based biomass shows that using the same land for biomass to replace coal would yield five to ten times the emissions savings of growing biomass for liquid transport fuel.[159]

Conclusion: car culture vs. sustainability & human rights

Agrofuels expansion is not about solving climate change. It is about keeping cars on the road. Humanity needs solutions to climate change that enable people to live decent lives. Agrofuels allow those who can afford it to maintain their lifestyles at the expense of the millions who will not be able to afford food, of the communities who are squeezed off their land, at the expense of biodiversity and vital ecosystems. As American environmentalist Lester Brown described it, agrofuels are a 'smash and grab raid by the 800 million people who own automobiles, on the fields that grow food for the planet's three billion poor'.[160]

Scores: First Generation Agrofuels

Effectiveness at reducing greenhouse gas emissions: 1/10
Living up to the hype (science to spin ratio): 1:8
Democratic ownership and control: 2/10 – increased power of global agri-business offsets the limited opportunities for local production and use
Social justice: 1/10 – food price rises, land-grabbing, workers' rights
Sustainability: 1/10 – expansion is reliant on land conversion and damage to ecosystems
Scalability: 1/10

Carbon capture and storage / 'Clean' fossil fuels

Carbon capture and storage (CCS) is a major departure from other climate mitigation strategies as, rather than stopping a damaging activity or replacing fossil fuel use, it allows current activities to continue but captures the carbon emissions and buries them under the ground. CCS is a major plank of the European Union's strategy of moving towards 'sustainable fossil fuels'.[161] The CCS approach fails to recognise the wider systemic problem of overconsumption or deal with the other impacts of that overconsumption.

CCS is also closely associated with enhanced oil recovery (EOR), where the carbon dioxide pumped into ageing oil wells makes more oil available, which in turn produces more emissions when burned. Companies' duty to pursue profit above other concerns means CCS would be used with EOR wherever possible unless this is actively prevented.

For some environmentalists, CCS is a lesser evil than either nuclear power or devastating climate change. But could it work in time, and what are its side effects?

How the technology works

CCS is a system for capturing, transporting and storing carbon dioxide from large emitters. Most of the work is currently going into applying this technology to power stations but it could theoretically be used to capture the emissions from other large single-site emitters such as cement, iron or steel works and oil and gas refineries.[162]

Capture

There are three main methods for capturing carbon dioxide from a power station being put forward:[163]

Pre-combustion capture involves gasifying the fuel to separate it into hydrogen and carbon dioxide. The hydrogen is lighter so the two gases are easily separated and the carbon dioxide stream is relatively pure. The hydrogen is then burnt to power a generator and produce electricity.

Post-combustion capture chemically 'scrubs' the carbon dioxide from the mixture of gases produced during the combustion process. This has the advantage that it can be retrofitted to existing plant.

Oxyfuel combustion involves burning the fuel in oxygen rather than air, meaning that the exhaust gas consists primarily of water vapour and carbon dioxide. When the water vapour is condensed out, the remaining carbon dioxide gas can be captured.

Each of these options requires a large amount of extra energy - 10-40% more depending on the technique used.[164] This energy is used to compress the carbon dioxide for transportation and in the capture process - either to gasify the fuel, to 'scrub' the flue gas of carbon dioxide, or to extract oxygen from the air.

Transport

The most effective means of transporting the captured carbon dioxide is via pipelines, similar to the gas pipeline infrastructure that already exists. Tankers could also be used but would be more expensive and less efficient.[165]

Storage

There are three main options for storage of captured carbon dioxide:[166]

> **Geological storage** in depleted oil and gas fields, saline aquifers (geological formations containing undrinkable water with high salt contents), or unmineable coal beds.
>
> **Ocean storage** either by dissolving the carbon dioxide in ocean water or pumping it deep down to the ocean floor. Below a certain depth and pressure in the deep ocean, carbon dioxide liquefies and becomes denser than water so in theory should form a stable lake on the ocean floor.
>
> **Mineral sequestration** in which the carbon dioxide reacts with quicklime (calcium oxide), to form limestone (calcium carbonate). The IPCC has estimated that a power plant equipped with CCS using mineral storage would need 60-180% more energy than a power plant without CCS.[167]

Schematic diagram of possible CCS systems

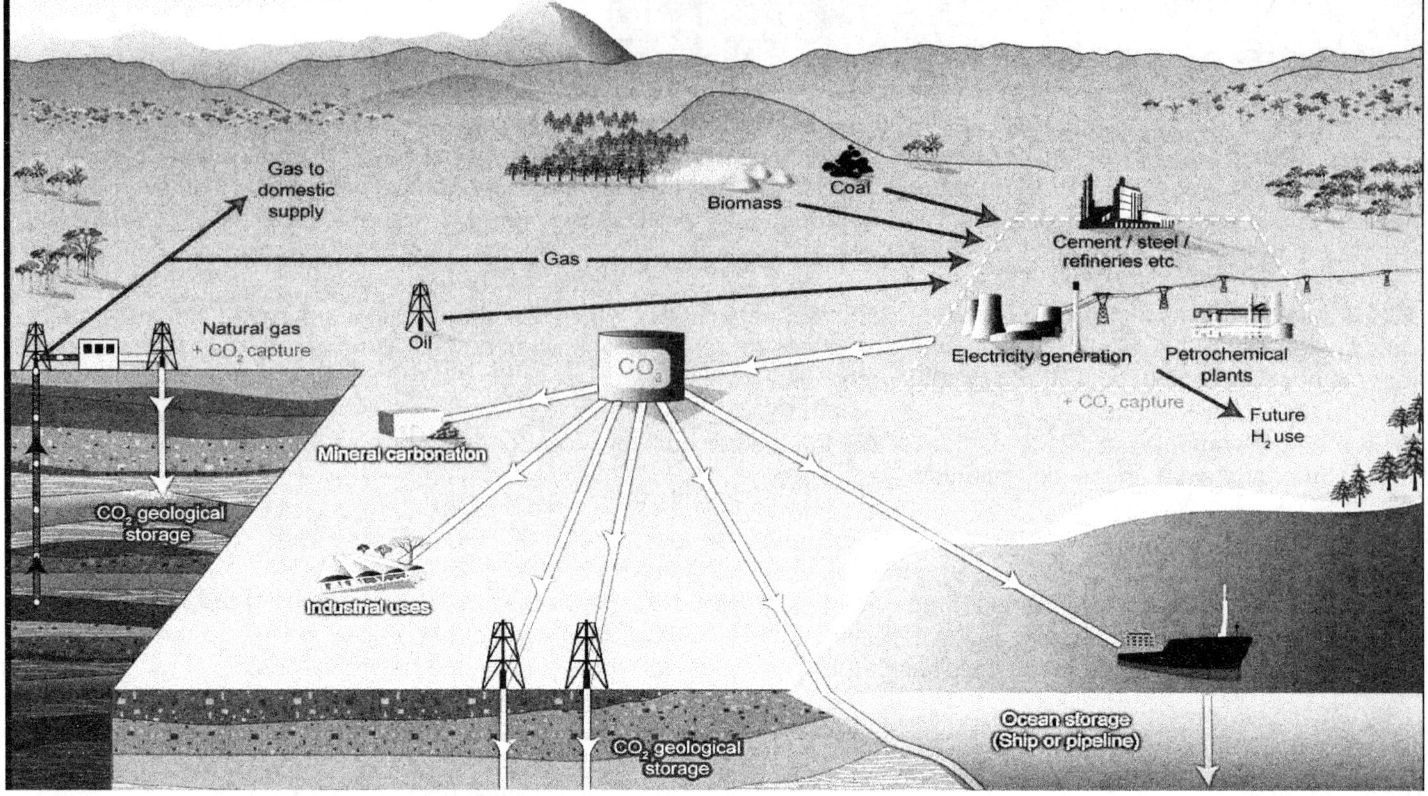

Carbon Capture & Storage Current Position

Status of the technology

Aspects of this technology already exist to various levels of maturity as carbon dioxide is already used by the oil industry to enable exploitation of oil reserves that would otherwise stay in the ground. This technique is known as Enhanced Oil Recovery (EOR). Carbon dioxide pumped into ageing oil fields dissolves the oil, making it flow more easily so that it can be pumped. Norwegian oil company Statoil has been pumping carbon dioxide into the Sleipner field in the North Sea since 1996,[168] this is simply for storage not for EOR. This carbon dioxide has been removed from gas deposits that have too high a carbon dioxide content to be allowed onto the market.[169]

At the capture stage, both pre and post combustion technologies are used in some commercial projects. Pre-combustion capture is used in fertiliser manufacturing.[170] Post combustion capture is used in the natural gas processing industry.[171]Applying this technology to power plants would require further development and demonstration projects. Oxyfuel combustion is at the demonstration phase.[172] Pipeline transport of carbon dioxide is an existing technology (in the USA, over 2,500 km of pipelines transport more than 40 Mt/carbon dioxide per year).[173] Of the storage options, the geology of oil and gas fields well understood because of the experience of the oil industry, however more research is needed to confirm whether these structures are suitable to store carbon dioxide for thousands of years.[174] Less is known about storage in saline aquifers.[175] Another proposed use is to store carbon dioxide in coal seams to enable the extraction of methane; this technology is also at a demonstration phase.[176] Ocean storage and mineral carbonation are both at a research phase.[177] Systems for injecting carbon dioxide and monitoring storage need further development, including adaptation from existing applications used by the oil industry.[178] There is no method yet for determining how much can be stored and how to tell if things are going wrong.[179]

The IPCC suggests that from well selected storage sites, storage rates are likely to exceed 99% over 100 years.[180] Leakage is possible from sites in two ways, either large scale leaks during transportation or injection,[181] or gradual seepage over time from abandoned oil wells or damage to the geological structure as a result of earlier oil and gas exploration.[182] Potential sites would have to be well surveyed to minimise the possibility of seepage.

Although the technologies needed for various aspects of a CCS system exist to varying degrees of development, there are no examples of a fully integrated CCS system in existence.[183]

Support for the technology

Many governments are currently supporting the development of carbon capture and storage, including demonstration projects in the USA, UK, Norway, EU, Australia and China.

Developing a fully integrated CCS system would require numerous demonstration projects adapting various systems for capture with different designs of power plant and testing out options for storage. These programmes would need to be run for a number of years in order for the technology to be properly demonstrated. The UK government anticipates that this would take at least 15 years.[184] The EU estimates that CCS could be commercially available in 2020[185] and that 12 demonstration plants of 300MW would be required costing at least €5bn (£4bn).[186] The UK has announced a competition for a demonstration project to apply post-combustion CCS to either a new or old power plant.[187] The USA's main demonstration programme, FutureGen, will not be operating until 2015.[188] The UK government's scenario envisages emissions reductions from CCS not starting until 2020, and increasing from a very low base of initially saving 300,000 tonnes of carbon per year.[189] For comparison, the UK's largest coal plant, Drax, alone produces over 20 million tonnes of carbon dioxide each year.[190]

Major players

Companies likely to bid for UK CCS demonstration competition:
Powerfuels, E.On, Scottish Power, Scottish and Southern Energy, Centrica, Progressive Energy, Conoco Philips, RWE

Members of Futuregen:
American Electric Power, Anglo American, BHP Billiton, China Huaneng Group, CONSOL Energy, E.ON, Foundation Coal, Luminant, Peabody Energy, PPL Energy Services, Rio Tinto Energy America, Southern Company Services, Inc, Xstrata Coal

Carbon Capture & Storage Issues

Delaying action on climate change

CCS is only useful as a solution to climate change if it can be used now and scaled up. A study by MIT suggests CCS will not be operating on a commercial scale until 2030. Shell says it will not be in widespread use until 2050. But even taking the optimistic projections of various governments, if CCS will not be available on a large scale until at least 2020 it cannot contribute to peaking global greenhouse gas emissions by 2015 or to managing the energy descent in time to avoid devastating climate change. Because waiting for its introduction is being used as an excuse to keep burning coal, CCS will not so much smooth the transition to a low carbon economy as delay it.

The optimistic timescales already look like they are beginning to slip. Despite statements from senior executives saying, 'developing commercially viable CCS should be a priority for companies and governments all over the world', Shell, along with its partner Statoil, ditched a CCS demonstration project in Norway which would have enabled EOR from their Draugen and Heidrun oil fields due to lack of profitability.

The US government's FutureGen project, a joint initiative between the government and the coal industry, is also floundering. Having originally planned a single demonstration plant and completed a tendering process to determine the location, the Department of Energy announced in January 2008, just six weeks after the location of the project was announced, that the project is being restructured and the money will instead be shared between a number of projects.[191] This has put the date the demonstration plants are supposed to begin operating back from 2013[192] to 2015.

For all the spin, governments and companies have not been prepared to put their money where their mouth is on CCS, waiting for the carbon market to kick in and make the projects economically viable. The projected dates of 2016 and 2020 may well be overoptimistic. When will real emissions reduction action begin? Other technologies are available now.

The potential to reduce carbon emissions is over-hyped

The IPCC estimates that by 2050 20-40% of global fossil fuel emissions would be technically suitable for capture including 30-60% of emissions from electricity production and 30-40% of emissions from industry.[193] Actual emissions captured would inevitably be less than this, and there would be additional emissions resulting from the processes of CCS which can itself consume up to 40% of the energy produced by a power station depending on the method used.[194] In addition, energy used to extract and transport coal takes up to quarter of the energy coal produces at the power station. These emissions cannot be captured. The most comprehensive assessment, made by Peter Viebhan of the German Aerospace Centre suggests that CCS can reduce greenhouse gas emissions from coal power stations by little more than two thirds.[195] This is less than the deep cuts which scientists are saying need to be made.

Justifying the expansion of coal

Choosing CCS means choosing coal. While CCS could theoretically be used for capturing emissions from a range of large point sources of carbon dioxide, most of the interest is in capture from coal power stations. Coal is a much more abundant resource than oil and gas (reserves exist to last up to 147 years at current rates of production according to the World Coal Institute[196]), coal reserves are widely geographically spread, and coal is cheap. CCS is expensive, so to become competitive the fuel needs to be cheap.

CCS is already being used as an excuse for the expansion of coal. For 20 years, no new coal power stations have been built in the UK.[197] With many of the UK's coal power stations coming to the end of their life and expected to close by 2015,[198] the coal industry is using the possibility of future CCS technology to push the government into granting permission for a new generation of coal plants. German energy company E.ON is planning two new 800 MW coal power stations at the site of its current plant at Kingsnorth in Kent.[199] The plant will emit more than 8 million tons of carbon dioxide a year.[200] Nationally there are another six coal power stations in the pipeline.[201] It is assumed that new plants will be built 'capture ready', so that CCS technology could potentially be added if it becomes commercially viable.

In practice, this means little more than designating space for a future carbon capture plant at the site. It is unclear whether the government will require all plants to be capture ready in any case.[202] Emails between the Department for Business Enterprise and Regulatory Reform (BERR) and Kingsnorth's owner, E.On, have revealed that E.On vetoed BERR's suggestion of including CCS in the conditions for the plant.[203] Since then, E.On have announced that Kingsnorth will be entered into the government's competition for funding for a CCS demonstration plant.[204] If they do not win will they cancel their plans?

The new Kingsnorth plants are expected to open in 2013. In the best case scenario, where the new plants would be adapted for carbon capture straight away when the technology becomes available and economically competitive (the technology may be available and tested as a complete system by 2020, but it could take a decade for the system to be applied to all existing plant)[205] there will be a significant overlap where the new plants are emitting huge quantities of carbon dioxide with no CCS in place. Building new coal plants in Kent is also problematic. Kingsnorth power station is a long way from potential storage sites in the North Sea oil fields. If coal and CCS are to be used then it makes more sense to concentrate plants further north than to build pipeline infrastructure across the country. The government has acknowledged this, saying, 'we will need to recognise that for some projects the scope for CCS (eg because of geographical location or other technical limitations) may be limited'.[206] In addition, retrofitting existing plants is much less economical than building new plants with carbon capture once the technology is ready. The EU estimates the cost of retrofitting 'capture ready' plants to be €600,000 – 700,000 (£475,000-554,000) per megawatt of capacity.[207] So building new coal plants under the excuse of future CCS increases medium term emissions and the eventual cost of emissions reductions.

Expanding coal use means more mining. In 2007, work began at the new Ffos-y-Fran open cast mine in Merthyr Tydfil, Wales, one of the largest in Europe.[208] The mine covers 400 hectares[209] with the edge of the site only 36 metres from people's homes[210] and is dubbed a 'land reclamation scheme' by the mining company, Miller Argent.[211] It will excavate 11 million tonnes of coal[212] resulting in 30 million tonnes of carbon dioxide emissions.[213] Ten new coal mines were approved by the Labour government in 2006.[214] The UK government's commitment to 'securing the long term future of the coal industry',[215] seems to outrank reducing emissions.

Sustainability impacts of coal

CCS means burning more coal than is used in a conventional coal plant, so bar the impact of the carbon dioxide which is captured, all the other unsustainable impacts of coal are multiplied. Coal is the dirtiest of all fuels, from its mining to disposal of waste products. It releases toxic pollutants into the air, water and land, and destroys habitats. The coal industry claims that with 'clean coal' technology these impacts are reduced, but even with the most cutting edge technology harmful emissions from coal plants are not eliminated, and in most cases they remain higher than for other forms of electricity generation. These technologies are expensive and with the expense of CCS already added to the cost of electricity generation it is highly unlikely that they would ever be universally applied.

Sustainability impacts of carbon storage

The main impact of geological storage of carbon dioxide is from leakage. Large scale leaks would have a global impact on the climate. Can anyone be sure that significant leaks would not happen because of human error or technical faults? The effect of seepage of carbon dioxide on ecosystems is little understood. In places where carbon dioxide leaks naturally from the earth ecosystems are adapted to it. There is a danger of significant ecosystem disruption if carbon dioxide seeps into an ecosystem that is not adapted to it. Carbon dioxide dissolved in water creates a mild acid, and seepage could pollute drinking water sources or contribute to ocean acidification. There is a concern that companies wanting to minimise the cost of transportation of the carbon dioxide will use less than perfect sites that are closer to the sources of emissions. If this technology is used, any leaks would have to be found and remedied, yet the technology to ensure this does not exist. There are significant legal questions around who will be responsible for long term storage – similar questions around responsibility for nuclear waste have proven intractable.

With ocean storage of carbon dioxide there is a distinct likelihood that this technology would contribute to the acidification of the oceans caused by climate change, which is already a grave concern. There is also little chance of effectively monitoring ocean storage. This approach is extremely risky and should not be pursued.

Pumping fossil fuels that would otherwise stay in the ground

The close association between CCS and enhanced recovery of oil, or methane from coal beds, is one key incentive motivating CCS' backers. Oil and gas reserves which are already accessible contain sufficient carbon to cause devastating climate change if burned. Accessing more fossil fuels does not make sense as part of a system for emissions reductions.

In 2005, BP sought government support for a CCS demonstration plant at Peterhead in Scotland. The project, a £500 million pre-combustion capture natural gas plant, would have been used for enhanced oil recovery (EOR) at BP's Miller oil field.[216] However, government support was not received and the project was cancelled. If it had gone ahead, BP claimed that 26 million tonnes of carbon dioxide would have been sequestered.[217] The carbon dioxide would have been used to pump an extra 40 million barrels of oil. Based on the emissions from an average barrel of oil, these 40 million barrels would have produced 12.68 million tonnes of carbon dioxide.[218] If captured carbon dioxide is used for EOR then it is by no means 'carbon free' energy as companies like BP claim.

Conclusion

Its proponents claim that CCS offers a route to gradually wean ourselves off fossil fuels while rapidly cutting carbon emissions. However, this thinking prolongs business-as-usual, preserving the dominance of the fossil fuel industries and diverting attention and investment from more sustainable methods of emissions reduction.

For CCS to be part of a sustainable solution, it would have to be used as a strictly interim measure, not longer than the lifespan of a power station, to enable gradual transition from high emission fossil fuel use, building up of renewables capacity and changes to lifestyles and economies to use less energy. It would not be used to enhance recovery of fossil fuels, to perpetuate the dominance of the fossil fuel industries, to justify the expansion of the coal industry while CCS technology is not available, or to divert funding away from renewables. Since this scenario is a long way from reality, and is unlikely to emerge in time to be useful, CCS will not be an effective solution to climate change.

Scores: Carbon Capture & Storage

Effectiveness at reducing greenhouse gas emissions: 3/10 – unlikely to be available in time, and emissions reductions are offset by the effect of enhanced oil recovery
Living up to the hype (science to spin ratio): 1:7
Democratic ownership and control: 1/10 – perpetuates dominance by big energy corporations
Social justice: 3/10 – no new side effects, but perpetuates injustices of the coal industry
Sustainability: 3/10 – absolute sustainability limited by coal reserves, also mining has negative impacts on ecosystems
Scalability: 4/10 - only applicable to very large emissions sources.

Nuclear Power

The arguments around nuclear power are well rehearsed elsewhere,[219] but this report would not be complete without an analysis of whether nuclear power is a useful technology in the fight against climate change and what the impacts would be of a new generation of nuclear power plants.

The nuclear industry has jumped on the climate change issue as a last ditch attempt to survive in the face of long-term public opposition, cost escalation and the intractable issue of what to do with nuclear waste. The industry has orchestrated a well-executed spin campaign and has succeeded in putting 'the nuclear option' back on the table.[220]

The arguments used are that nuclear power is 'clean' and has low emissions of carbon dioxide, the uranium can be sourced from friendly countries such as Australia and Canada without the need to deal with 'unstable' Middle Eastern countries, that the supply is constant (unlike wind or solar), and that it will fill the gap in energy supply that some commentators have predicted as likely to happen when the current generation of coal and nuclear electricity plants come to the end of their life between 2015 and 2030. However, nuclear power is not carbon neutral, would not be on-stream in time to plug the energy gap or avoid dangerous climate change, and has huge environmental impacts, not least from the huge quantities of radioactive waste created, which no country has yet developed an effective way of dealing with.

Nuclear Power Current Position

Status of technology

Nuclear power is a mature technology, and significant further developments in efficiency and cost-effectiveness seem unlikely. However, it is not necessarily dependable since most plans are for new designs of reactor which have not been built before.

Proportion of current usage

18% of UK electricity generation.[221]

Support being received

There have been no nuclear power stations built in the UK for 20 years, but a new generation of nuclear power stations was announced in the 2006 Energy Review.[222] The government was forced to review this decision after Greenpeace successfully challenged it in the High Court because the policy had not been adequately consulted upon. The judge ruled that the government's consultation had been 'misleading', 'seriously flawed' and 'procedurally unfair'.[223]

After further consultation, on 10 January 2008, Secretary of State John Hutton announced the government's resolution to go ahead with plans for new nuclear power stations.[224]. Ten new reactors are expected.[225]

Major players[226]

Electricity generators:
British Energy, EDF Energy, E.On, RWE

Construction:
AMEC, Balfour Beatty, Carillion, Sir Robert McAlpine

Nuclear engineering and design:
Areva, EPR, Westinghouse, Atomic Energy of Canada

Nuclear fuel and services:
British Nuclear Fuels Ltd (BNFL), Unreco

Nuclear Power Issues

Nuclear power is not a solution to climate change

Nuclear power is presented as clean energy because no carbon dioxide is emitted during the electricity generation process. But other energy is required for every other stage in the process including the mining, milling and transportation of the uranium, the construction and decommissioning of the power plants, and the reprocessing, storage and (eventual) disposal of nuclear waste. At present, most of this energy comes from fossil fuels.

Life-cycle assessments of the greenhouse gas impacts of nuclear power have produced widely varying and probably incomplete results.[227] It is difficult to incorporate an assessment of emissions from waste disposal without knowing how waste will be disposed of, so not all studies incorporate this essential component.

But what are the likely emissions levels?

The UK government estimates the carbon dioxide equivalent emissions per kilowatt hour (kWh) of electricity produced from nuclear power at 7-22g.[228] Natural gas is responsible for 450g CO2 e/kWh.[229] A study by Sydney University's Integrated Sustainability Analysis suggests that the figure is more like 10-130g CO2 e/kWh with an average of 65g CO2 e/kWh.[230] Another study by Storm Van Leewen calculates that taking into account best practice and the reality of errors and problems in the nuclear cycle, emissions of carbon dioxide are in the range of 88-134g CO2e/kWh[231]. On this basis, emissions are certainly less than gas, but likely to be far from carbon-neutral.

All of these studies assume that the ore used is as high-grade and easy to mine as that in current use. However, as the easily mineable and high grade uranium runs out lower grades and more energy intensive mining techniques would have to be used, significantly increasing emissions.

The Sustainable Development Commission, the government's main advisors on sustainability issues, says that in the most optimistic scenario, a programme of new nuclear power plants could reduce carbon dioxide emissions in the UK by 4-8% by 2035.[232] This does not, however, take into account the delays associated with the nuclear industry. The new 'third generation' reactor technologies have not been built anywhere in the world. The first under construction, at Olkiluoto in Finland, was supposed to be a showcase for the simplicity of the new design, but has been beset by delays and cost overruns.[233] The reactor is four years from completion and has already seen cost overruns of €1.5 billion (£1.2bn). It was 24-30 months behind schedule after just 27 months of construction.[234] In fact no nuclear power station has ever been built in the UK without delays and cost overruns and many have also failed to generate as much electricity as promised.[235] Even with the Sustainable Development Commission's optimistic figures, new nuclear power plants would take at least a decade to build, with the first not on stream for at least fifteen years. Even aside from the other issues, this is too little too late to help.

Outside the UK, there are even greater issues with increased use of nuclear power. Nuclear power is, of course, closely associated with development of nuclear weapons, including the possibility of terrorist 'dirty bombs'. Even if other issues could be ignored, this means that nuclear would never be suitable for use in politically unstable countries, or countries facing economic or other crises which could lead to lapses in safety standards and increased likelihood of devastating accidents. As climate change worsens, the number of countries facing such crises will increase - new nuclear power stations make this prospect even more terrifying.

Waste

No country has solved the issue of how to safely dispose of nuclear waste. All the high-level waste produced in the past 60 years of nuclear power generation is sitting in stores waiting for an answer. The answer will need to last for thousands of years.

If the problem of the existing stockpile of tens of thousands of tonnes of nuclear waste is not tackled during the era of cheap fossil fuels then it seems unlikely that will ever be solved. Without fossil fuels, how is it proposed to pack the waste away into secure canisters lined with lead, steel and pure electrolytic copper using remotely-controlled robots and dump them in the geologic storage repositories that are yet to be established?[236]

Reprocessing nuclear waste to produce more fuel may be able to reduce the need for new mined uranium (although not necessarily - the recent revelations about the UK's mixed-oxide fuel reprocessing plant at Thorp, which has produced almost nothing in its first six years of operation, suggest that this technology is still problematic)[237]. But reprocessing increases the amount of waste to be dealt with as large quantities of materials are rendered radioactive.
The UK's proposed underground waste facility is waiting for a community to volunteer to host it. It is predicted that

it will take 40 years to complete once the design is ready – after a community has stepped forward.[238] It cannot be proved whether underground burial will totally prevent the waste from leaking back into the environment, polluting water supplies and the food chain.

The creation of nuclear waste is inseparable from the use of this technology. It exposes future generations to enormous danger, for the sake of short-term political gain and corporate profit.

Impacts of uranium mining

Uranium is mainly mined in vast open-cast pits. In some hard-to-reach seams uranium is removed through in situ leaching, where sulphuric acid, nitrous acid and ammonia are injected into the seam and pumped up again years later.[239] This leaves a huge amount of radioactive waste in the environment. The ore is then milled. Typical grade uranium ore requires1000 tonnes of rock to be ground up to produce one tonne of useful fuel. The other 999 tonnes of rock is radioactive indefinitely and is left in the environment where its radioactive products are free to be leached out.[240] Treating this waste is regarded as ideal, not as best practice, and would take four times the energy currently used to mine the uranium.

Cost

The government has said that it will only allow the private sector to build new nuclear power stations without government subsidy. However, assistance given in the past has turned into huge subsidies and long-term government financial commitments including a bail-out of British Energy in 2002 which has left the government with liabilities amounting to £5.1billion.[241] The industry claims that, because of higher fossil fuel prices and carbon pricing, nuclear power is now economically competitive. However, the economics of nuclear power are highly uncertain, particularly with the new generation designs that have never been built before. UBS Investment Research say that 'endorsing new nuclear is... a potentially courageous 60-year bet on fuel prices, discount rates and promised efficiency gains...'[242] HSBC says that new nuclear would be a 'difficult pill to swallow for equity investors'.[243]

Government subsidies and sweeteners for nuclear power are often hidden from the public. Any company that builds a new nuclear plant would be liable for its waste and decommissioning costs. However, the government may choose to place a cap on this or to transfer liability to the Nuclear Decommissioning Authority as has happened in the past. Other possible support the government may give includes underwriting commercial loans for new build, guarantees on cost overruns or tax breaks. Whatever deals the government strikes now with the nuclear industry, if anything goes wrong the taxpayer is likely to pick up the tab.

The estimated £62.7 billion costs for the UK's existing nuclear waste and decommissioning liabilities[244] will come from public money.

The nuclear industry is thought to have demanded that the government guarantee a minimum carbon price for the life of the new stations.[245] If the carbon price were to collapse the taxpayer would be obliged to make up the difference.

Diverting investment from renewables

The huge financial commitment to nuclear the UK government is planning would inevitably divert investment away from other, potentially more effective, ways of tackling greenhouse gas emissions from the energy sector.
Amory Lovins from the US Rocky Mountain Institute has calculated that 'Each dollar invested in electric efficiency displaces nearly seven times as much carbon dioxide as a dollar invested in nuclear power.'[246]

In announcing the results of the UK government's Energy Review 2003, Patricia Hewitt, then Trade and Industry Secretary, admitted that investment in nuclear would come at the expense of renewables, saying, 'It would have been foolish to announce ... that we would embark on a new generation of nuclear power stations because that would have guaranteed that we would not make the necessary investment and effort in both energy efficiency and in renewables. That is why we are not going to build a new generation of nuclear power stations now.'[247]

It is clear that 'the necessary investment and effort in both energy efficiency and in renewables' has not been made in the intervening years, so why has nuclear suddenly become essential to the government's plans?

Flooding and the impacts of climate change

Nuclear power generation requires huge quantities of water, which is why all of the UK's nuclear power plants are by the sea. These locations are liable to flooding due to sea level rise and storm surges, which are made more likely by climate change. Many existing and proposed nuclear sites should not be used for this reason according to researchers from the Flood Hazard Research Centre.[248] The Drigg facility at Sellafield, where intermediate level nuclear waste is stored, is likely to be destroyed due to rising sea levels over the next 500 years – well before the material it contains loses its radioactivity.[249]

Conclusion

For the UK government and the corporations involved, a programme of new nuclear power stations has two huge advantages as an approach to climate change – if one believes in the nuclear dream. Firstly, it allows the continuance of business-as-usual for the energy industry, with large centralised power stations owned by transnational corporations supplying a national grid – unlike wind, solar power or decentralised energy plans it requires no new tariffs or new industry structures. Secondly, it allows a semblance of business-as-usual to continue for consumers – politicians can pretend that with nuclear, no significant changes to our way of life will be necessary to prevent dangerous climate change.

Scores: Nuclear Power

Effectiveness at reducing greenhouse gas emissions: 4/10
Living up to the hype (science to spin ratio): 1:3
Democratic ownership and control: 1/10 – perpetuates dominance by big energy corporations
Social justice: 1/10 – waste must be dealt with indefinitely by future generations
Sustainability: 3/10 – absolute sustainability limited by fuel availability, waste problem gets worse the longer use continues
Scalability: 5/10 – not suitable for all global need

Nuclear Fusion - A Quick Word

A breakthrough in nuclear fusion technology has been 'about twenty years away' since around the 1950s, if the researchers are to be believed. Even if that were true today, it is not going to be developed and scaled up to be onstream in time to solve the current emergency. Despite this, €10 billion (£7.9bn) is going into the fusion research project for the International Thermonuclear Experimental Reactor (ITER)[250].

Solar Power

How it Works

Photovoltaics (PV)

In a photovoltaic panel, light energy induces an electric current in a semiconductor (usually silicon). PV functions anywhere there is light; it is effective even at low intensity but the greater the light intensity the greater the flow of electricity.

Solar thermal

Solar thermal technologies use sunlight to heat a fluid in a black container, either directly heating water or passing the heat to water via a heat exchanger. In the UK it is currently mostly used for domestic hot-water supply.

Concentrated solar power (CSP)

CSP uses intense solar radiation to heat water to turn a turbine and produce electricity. There are various different designs but each uses mirrors to focus the sun's energy on a receiver in which a oil is heated to temperatures of 400-1000°C[251], which in turn heats the water to drive a steam turbine. CSP requires intense sunlight, and has been used in California since 1985.[252] There is a great potential globally for CSP to be used, particularly in deserts. Electricity is only produced in the daytime and is difficult to store for use at night.

Passive solar

The use of building design to maximise heat gain from sunlight and minimise heat loss, eg by increasing window area on the south-facing side of a building. It is a design issue rather than a separate technology in its own right.

Solar Power Current Position

Status of the technology

Solar energy technologies are already functional and are being developed further. There is considerable scope for rapid engineering improvements and for major economies of scale in manufacturing and deployment. Price per kWh is falling rapidly.[253] Technological breakthroughs may be required to significantly increase energy storage potential (see below).

Proportion of current usage

Less than 0.1% of UK energy use (including PV and solar thermal).[254]

Support for the technology

The market for PV is greatest in Germany, the USA and Japan, with Germany taking a massive 58% share of the global market. This is due to government support for solar energy including `feed in' tariffs where users are paid a fixed rate for electricity returned to the grid.

In the UK, the Major Photovoltaics Demonstration Programme (PVMDP) which gave grants to install PV including to households, schools and public buildings, has closed and been replaced by the Low Carbon Buildings Programme covering a range of domestic low carbon technologies and energy efficiency measures.

New CSP plants are underway in Spain and the USA with potential markets in China. The high level of investment needed for large scale energy production is the main block to expansion of this technology.[255]

Major players

Photovoltaics Manufacturers:
Sharp, Kyocera, BP Solar, Avancis, Isofoton, SunTech, Siemens, Mitsubishi, Sanyo, RWE Schott Solar, Nanosolar, SolarWorld

Photovoltaics Installation:
British Gas – Interestingly, big energy companies such as British Gas are moving into the solar installation business which was previously the preserve of small companies. Their size means they can sometimes out-compete smaller companies, and British Gas is also going into partnership with local authorities and the government to offer incentives to customers installing with them.

Concentrated solar power:
Ausra, Flagsol, Acciona, Abengoa, Schott, Solar Millennium

Solar Power Issues

Price

PV is still very expensive. It currently takes three to eight years for solar panels to produce the amount of energy it took to make them, but this is predicted to drop to two to four years.[256] Solar panels currently have a 25-30 year life expectancy, but they only pay for themselves after 25-35 years.[257] Industry claims that costs are coming down, falling roughly 20% every time the industry doubles global manufacturing capacity, which is every two years.[258] Theoretically, solar also becomes more competitive when the full cost of emitting carbon is taken into account, but see the section on carbon trading in Chapter 3 for why this doesn't work in practice.

Encouraging take-up

Despite the fact that solar is the biggest area in cleantech investing, and even with projected increases in efficiency through new designs, it is hard to see that there will be a major transformative shift to solar in the UK without government schemes such as a 'feed in tariff' scheme similar to that introduced in Germany.[259] In that scheme, small-scale renewable generation is encouraged through a guaranteed preferential tariff for up to 30 years. Other EU countries offer schemes based on similar principles. The Stern Review considers feed-in tariffs to be an effective way of encouraging the deployment of renewable power generation.[260] One issue with this that in the UK, energy demand is highest in winter when light intensity is lower. Domestic users will sell to the grid when demand is low and buy from it when demand is high. So attempts to bring in a feed in tariff scheme would doubtless be resisted by powerful electricity companies for whom the economics of the scheme are unattractive.

It may be that large scale investment in solar is not the best option for UK public money when other renewable energy options are cheaper now, though the technology certainly has a part to play for off-grid and decentralised energy generation (see below). In other countries with greater solar resources and/or less developed distribution grids these issues may not apply.

Storage

All solar technologies work only when there is light – concentrated solar and solar thermal generally require intense direct sunlight. While hot water from solar thermal can be stored for hours or even days, storage for electricity is more difficult and less efficient, particularly on a large scale where few technologies currently exist. One idea is to use the excess electricity to pump compressed air into underground pipes and to release the air to turn a turbine at times of peak demand,[261] but technologies such as this are in their infancy.

Manufacture and disposal

Some solar technologies use substances which are toxic or carcinogenic or cause harmful environmental impacts when disposed of. Substances which may be released in multicrystalline silicon PV module production are fluorine, chlorine, nitrate, isopropanol, sulphur dioxide, carbon dioxide, respirable silica particles and solvents.[262] Levels of emissions of damaging substances are orders of magnitude smaller than the corresponding emissions of a coal plant. However some, such as silicon dioxide, may nevertheless be considered rather high for a technology with is considered sustainable.[263]

New 'thin film' solar technologies make use of new nanomaterials. Nanosolar have developed a nanoparticle ink which simply prints the semiconductor of a solar cell as though it were ink.[264] Development of nano-materials has outpaced research into their safety, and there is no regulation to ensure that the public and the environment is protected from any adverse effects.[265] The use of nanoparticles is particularly of concern as waste produced during the manufacturing process will be released into the environment causing unknown harm. Other new technologies use highly toxic substances such as cadmium.

Resource depletion may become an issue for some materials. For example, the world's entire current annual silver production would be required to achieve around a 20% contribution to world electricity supply using current solar technologies.[266] Other than aluminium frames, few of the materials from solar cells can currently be recycled.

Conclusion

Solar is one of the emissions-reducing technologies with the greatest potential for scaling-up to provide a significant portion of global energy needs. It also has the advantages of not needing significant new infrastructure and of being usable in isolated areas, both of which make it suitable for Majority World use. However, take-up will be hampered unless storage technologies can be developed.

In addition, solar is less attractive to traditional oil or power generation companies which have focused on ongoing sales of electricity or fuel, whereas solar mostly requires one-off sales of equipment. While there is undoubtedly money to be made in the coming decades by selling solar panels, margins are likely to be lower in the long term than for selling electricity or fuel.

Scores: Solar Power

Effectiveness at reducing greenhouse gas emissions: 8/10 - embodied energy of panels means not carbon neutral
Living up to the hype (science to spin ratio): 1:3
Democratic ownership and control: 7/10 – only manufacture requires centralisation and high technology - use can be very decentralised.
Social justice: 7/10 – no new negative side-effects and potentially allows extension of electricity availability into off-grid areas. Possibility of energy 'colonialism' (see 'Extending the grid') could reduce this figure.
Sustainability: 6/10 – may be some problems with raw materials availability and impacts of manufacturing
Scalability: 7/10 – could provide for a large proportion of global use, storage remains an issue.

Wind Power

How it Works

Wind turns blades connected to a turbine, generating electricity. Higher wind speeds generate more electricity, though most turbines shut down in very high winds.

Wind Power Current Position

Status of the technology

Various types and sizes of wind turbines are in use and the technology is developing rapidly. They range in size from micro turbines providing energy for individual homes, to 80m offshore turbines each generating 3MW of electricity.[267] Price per kWh continues to fall. Deployment is also advancing rapidly. As with solar power, technological breakthroughs may be required to significantly increase energy storage potential.

Proportion of current usage

Just over 1% of UK electricity generation in 2006 – this figure has more than doubled in two years.

Support for the technology

The global market for wind power has been expanding faster than any other source of renewable energy.[268] The largest markets for wind energy in 2005 were Germany, Spain, the USA, India and Denmark.[269] In Germany, premium tariffs for wind power encouraged investment.[270]

Investment in wind energy is encouraged by the UK government through the Renewables Obligation which requires power suppliers to buy a portion of the electricity they supply from renewables. The current level is 7.9% for 2007/08 rising to 15.4% by 2015/16.[271] Another incentive is the Climate Change Levy which taxes businesses and public institutions using energy derived from non renewable sources.[272]

Major players

Manufacturers:
Vestas, Nordex, Siemens, Clipper, GE, Enercon

Large UK Windfarm Owners and Developers:
Falck, Scottish and Southern, RWE Npower, Scottish Power, E.On, Centrica, Fred Olsen Renewables, Vattenfall

Wind Power Issues

Intermittency

Wind turbines produce electricity when the wind blows, not when you press a switch. The unpredictable nature of wind electricity production is known as intermittency. To keep the grid working demand has to exactly match supply. Pro-nuclear lobby groups have jumped on the intermittency issue to dampen interest in wind power.[273]

This issue could potentially be partly resolved by spreading wind farms around the country, or by having a more intelligently designed web of supply and demand management.[274] This, though, would require significant behavioural change (consumers would have to get used to not assuming round-the-clock access on demand) and some new technology of its own, eg to communicate with electricity users and encourage them to use high-energy devices only when supply is available. Methods for predicting wind flow are improving,[275] as are technologies for managing supply and demand.[276] As with solar, developments in storage technologies could also solve this problem.

Landscape impact

In the UK, some areas have seen considerable local opposition to the building of wind farms. While some of this opposition is purely aesthetic, and likely to evaporate as the real dangers of climate change become more widely known, and some is based on prejudice or encouragement by other vested interests, there are genuine objections to overcome. For example, there have been a number of wind farms proposed in the UK that have had significant negative environmental impacts[277] such as requiring large numbers of trees to be felled to reduce turbulence. Building wind farms on peat requires the peat to be drained and give up the large amounts of carbon it stores.[278] Building on birds' migration routes poses a risk to bird life.[279] All these issues can be avoided with foresight and environmental impact assessments, and still leave the UK with huge potential for developing wind power, although when commentators give estimates of the UK's potential wind resource these factors are often not taken into account.

Onshore or offshore

The wind blows strongest at sea. Offshore wind gives the potential for bigger turbines generating more electricity, but some of the energy is lost in transmission, and undersea cables are expensive. The UK has four offshore wind projects already generating electricity[280]. Six further wind farms are approved, and nineteen more are planned. Offshore wind avoids most of the aesthetic and some of the environmental objections faced by onshore projects but may still cause problems for birdlife.

Conclusion

Like solar, wind is one of the emissions-reducing technologies with the greatest potential for scaling-up to provide a significant portion of global energy needs, particularly in windy countries like the UK. Since larger turbines and turbines in exposed areas are more efficient, at present wind lends itself to deployment in large wind farms, though efficiency and costs improvements may eventually make small-scale wind viable for more than niche uses. As with solar, however, take-up will be hampered unless better storage technologies can be developed. Overcoming local objections to wind farms may require design improvements and ownership structures which give more advantage to local residents.

Scores: Wind Power

Effectiveness at reducing greenhouse gas emissions: 8/10 - embodied energy of turbines means not carbon neutral
Living up to the hype (science to spin ratio): 1:2
Democratic ownership and control: 5/10 – as with solar, could be very decentralised but current centralised deployment perpetuates dominance by big energy corporations
Social justice: 8/10 – no new negative side-effects and potentially allows extension of electricity availability into off-grid areas.
Sustainability: 8/10
Scalability: 7/10 – could provide for a large proportion of global use, storage remains an issue.

Water Power - hydroelectricity, tidal power, wave power

How it Works

Large-scale hydroelectric

A dam on a watercourse creates a reservoir. Water from the reservoir is gradually released through turbines to produce electricity. Some hydroelectric power stations can also be used to store energy. At times of low demand surplus power from the grid is used to pump water uphill into reservoirs to be released and generate electricity at peak demand.

Small-scale hydroelectric (micro-hydro)

Small plants use the flow of a river or stream to turn a turbine.

Tidal power

A tidal barrage works like a hydroelectric scheme, with a large barrier across an estuary which holds up the tidal flow (in or out) and releases water gradually, using the flow of the tide to turn a turbine or to push air through a pipe to a turbine. A similar system called tidal lagoons uses artificial pools which flood at high tide and empty on the ebb, using the flow of water to generate electricity.

Other tidal power designs work more like small-scale hydro, with turbines moving with the flow of the tide like underwater wind turbines.

Wave power

Several designs for exploiting wave power are being developed. Certain common elements can be observed, in particular, a wave power machine needs to resist the motion of the waves in order to generate power, ie part of the machine needs to move while another part remains still. Some designs include:

Oscillating water column: a partially submerged, hollow structure installed in the sea. It is open to the sea below the water line, enclosing a column of air on top of a column of water. Waves cause the water column to rise and fall, which in turn compresses and decompresses the air column. This trapped air is allowed to flow to and from the atmosphere via a turbine.

Buoyant moored device: this floats on or just below the surface of the water and is moored to the sea floor. In this type of device, the mooring is static and is arranged in such a way that the waves' motion will move only one part of the machine.

Hinged contour device: a long articulated tube, the size of several railway carriages, with hydraulic fluid in the joints. The resistance to the waves is created by the alternate motion of the waves, which raises and lowers different sections of the machine relative to each other, pushing the fluid in the joints through hydraulic pumps to generate electricity. [281]

Water Power Current Position

Status of the technologies

Large-scale hydro – mature technology, in use since 19th century. Large dams are still being built around the world, particularly in the global South.

Small-scale hydro – developing. Effective designs are available, but deployment is mostly slow. China currently has over 50% of the world's small hydro capacity and is continuing to promote its use in rural areas.[282]

Tidal power – experimental and developing. Some tidal power is in use, but new designs may be needed to develop technologies which are widely applicable and avoid adverse local environmental impacts. The world's first megawatt-scale commercial generator of the 'underwater wind turbine' design has been installed in Strangford Lough, Northern Ireland, and will permanently feed into the grid from August 2008 [399]

Wave power – experimental and developing. Engineering improvements would be needed to create an efficient and widely applicable technology.

Proportion of current usage

Large-scale hydro – approximately 1% of current UK electricity generation, mostly in Scotland. One 100MW dam is under construction at Glendoe. [283]

Others – less than 0.1%[284]

Support for the technologies

In the UK, expansion of large scale hydro is limited due to environmental concerns and the fact that most of the potential sites have already been used. Interest in large scale projects has shifted to tidal barrages, with construction and engineering companies lobbying for a tidal barrage on the Severn estuary.

Elsewhere in Europe, growth in hydroelectric schemes has been closely associated with growth in heavy industry, for example aluminium smelters which require huge amounts of cheap energy.

Major players

Large-scale hydro – In the UK the largest owner of hydroelectric systems is Scottish and Southern Energy.

International hydropower industry – Includes construction and engineering companies such as: Norsk Hydro, Impreglio, Voith Siemens, Halcrow, Alstom, Mott Macdonald.

Severn Barrage – The Severn Tidal Power Group (STGP) is a consortium of engineering and construction companies formed in 1981 to push for the construction of a Severn Barrage. Its members are Balfour Beatty, Taylor Woodrow, Sir Robert McAlpine and Alstom.

Water Power Issues

Large scale hydro

There are major social and environmental issues with large scale dams.[285] They gather silt behind the dam wall, block the flow of nutrients and wildlife downstream, fragment many of the world's major rivers and have displaced hundreds of millions of people. They are an environmental catastrophe in wildlife terms.

But is the energy produced from dams free from greenhouse gas emissions? No. When a dam is built and the land behind is flooded, the submerged plantlife decays, releasing stored carbon as either carbon dioxide or methane. For several years after the land is flooded the immersed vegetation gives off a huge pulse of methane. Even after it subsides, methane production continues indefinitely as seasonal drops in reservoir levels allow plants to grow which later get submerged.[286] Methane is 25 times stronger as a greenhouse gas than carbon dioxide.[287] Over a hundred years artificial reservoirs (about a quarter of which feed hydroelectric dams) will contribute about 7% of the global warming impact of all human activities.[288]

A study of the greenhouse gas emissions from the Curuá-Una dam in Brazil showed that, over than a decade after filling, cumulative emissions were nearly four times worse than if the same amount of electricity had been generated from burning oil.[289] The effect varies widely from dam to dam; it is much worse in tropical areas where plant growth is more vigorous, and in reservoirs of new dams (where the entire lake floor may consist of decaying plant matter).[290] However, studies in temperate areas show that the greenhouse gas contribution is still significant; at best it appears a dam gives one-tenth of the greenhouse effect of generating the same power from fossil fuels.[291]

However, decaying vegetation is not the only source of emissions from dams. The vast majority of dams are made from concrete and cement, sometimes millions of tonnes in a single dam. Manufacturing cement produces around 5-10% of global carbon dioxide emissions, at a rate of a tonne of carbon dioxide per tonne of cement[396]. So large dams may be responsible for millions of tonnes of carbon dioxide emissions before they are even filled.

The World Commission on Dams has said: 'there is no justification for claiming that hydro-electricity does not contribute significantly to global warming'.[292]

Small scale hydro

Small scale hydro projects, particularly systems running off the flow of a river or stream, avoid the negative environmental impacts of larger hydroelectric schemes. They also have the advantage of being suitable for use in isolated areas away from power grids. While small-scale hydro is only suitable for some sites, and output is not entirely consistent as it will vary seasonally depending on the flow of the watercourse, there is nevertheless scope for considerable further deployment and development with few side-effects.

Tidal power

Currently only three tidal barrages exist worldwide,[293] the largest on the Rance river in Brittany, France.[294] The main reason why the technology has not been used more worldwide is that it is hugely expensive, only a limited number of suitable sites exist, and because of the destruction of the ecosystems of estuaries.[295]

The 'underwater wind turbine' generators are very new technology and their potential impact on marine life is unknown. Just as early wind advocates didn't foresee the impact on birdlife, so tidal blades may have significant impact on marine life. The Strangford Lough generator's initial three months involve only operating during daylight hours and having a marine mammal observer onboard at all times, plus sonar to monitor seal movements.[399]

Npower are planning to install seven similar devices in a 10.5MW project off the coast of Anglesey, although due to the length of planning this wouldn't be in commission until 2011 at the earliest.[400]

The Severn Barrage

Since the 19th century there have been plans to build a tidal barrage across the Severn estuary as the site has the second greatest tidal range (difference in height between low and high tide) in the world, but none of the proposals have ever got off the ground. However, the UK government has indicated its support for a 'Severn Barrage'[296] following a report from the Sustainable Development Commission which supported the scheme on condition that it was publicly owned and money was put into a 'compensatory habitat' for threatened wildlife.[297]

A 16km barrage in the Severn Estuary could supply 4.4% of UK electricity demand (17TWh) generating electricity for over 120 years, with construction costs in the region of £15 billion.[298] However, the impacts on the local ecosystem are huge and irreversible. It would result in the loss of the mudflat, saltmarsh and sandflat habitats which are internationally protected as a site for wading birds.[299] The barrage would also have a significant adverse impact on fish populations as they get killed or injured in the turbines, their migration routes are blocked and the spawning grounds of many species would be beyond the barrier.[300] A further disadvantage is that the project would produce power in two massive 'pulses' each day, which though predictable would often be unaligned with consumer demand and potentially be difficult to accommodate on a grid based on intermittent renewables.[301]

The Environment Agency chief executive Barbara Young has slammed the proposed Severn Barrage scheme saying: 'A project to deliver 5% of the UK's energy at the price of wrecking valuable wildlife is not the way forward...If you wrote someone a note, you would not reach for the Mona Lisa to write it on.'[302] Friends of the Earth have proposed an alternative scheme using less intrusive 'tidal lagoons'. They claim that their scheme would be cost-effective and less ecologically damaging, protecting rare habitats.[303]

Wave Power

Waves are a powerful source of energy, but the energy is difficult to harness as any technology must withstand rough conditions and be able to generate from small as well as large waves. At present there is no reason to foresee significant side effects from increased deployment of wave power.

There are two wave power devices in the UK. Total capacity currently stands at 1.25 MW[304], with many new projects in the development stage including a 3MW Scottish Power project in Orkney. According to the government the UK has wave power levels that are among the highest in the world. Wave energy has the potential to provide as much renewable energy as the wind industry, but the development of wave technology is currently at the same level as the wind industry was 10 years ago.[305] However, the changes to the Renewables Obligation in 2009 will add an extra incentive to use wave power and tidal stream.

Biomass

How it Works

Biomass as an energy source refers to the burning of wood, other plant material or dung, either for direct use as heat or to generate electricity.

In this section we include the use of biogas – methane for fuel captured from the decomposition of plant and animal waste in the absence of air. The methane can be used either for heating or for generating electricity.

Landfill gas and incineration are dealt with separately – see below.

Biomass Current Position

Status of the technology

Burning biomass is humanity's oldest energy technology – the majority of biomass fuel use globally is still made up of traditional heating and cooking fires and stoves.

Proportion of current usage

14% global energy use, mostly traditional biomass. Biomass is globally the largest contributor to renewable energy production.[306]

Approximately 0.6% of UK energy use, including wood and other plant-based solid fuels and sewage gas.[307]

Support for the technology

Biomass is supported in the UK through the Renewables Obligation which requires electricity suppliers to use a set amount of renewable energy[308], and through action plans to encourage the growth of fuel crops and promotion of biomass through management of forestry. The government also operates grant schemes such as the £400 million Environmental Transformation Fund which supports a range of renewable technologies in the UK[309]. In 2008/09, £10 million of this fund will go towards capital and infrastructure grants for bio-energy.

Major players

EDF, Drax, Prenergy Power, E.On, SembCorp Utilities, RWE, ESD Biomass, Scottish Power, Renewable Fuels Ltd. Biomass UK.

Biomass Issues

Co-firing

The main use for biomass in large scale electricity production is in co-firing biomass with coal in power stations. This has been exploited in public relations campaigns by the coal-fired power generation industry to deflect criticism while climate change demands their closure. For example, Drax power station, the UK's largest coal-fired power plant, producing 7% of the UK's electricity, trumpets its co-firing even though, until this year, it represented less than 2% of its total production. Drax made a big play of using locally sourced fuel, saying that 'Drax Power has already made much progress in promoting local supply partnerships for future forestry and energy crops',[310] but has since been using olive pips transported from Italy as feedstock,[311] with the attendant transport emissions.

Co-firing with biomass also faces major problems of scalability. For example, if Drax were to achieve their short-term goal of producing 10% of its output from biomass, it would require 450,000 hectares of willow on 3-year rotation coppice.[312] On this basis, an area one-quarter the size of the UK would need to be planted with willow in order to produce 10% of the UK's electricity from co-fired biomass.

Scottish Power's Longannet and Cockenzie plants have introduced biomass requiring 12% of Scotland's total agricultural land.[313] This is to displace only 5 % of the company's coal requirement by 2013.[314] There are no figures currently available on levels of indirect carbon emissions due to land use for UK biomass.

Co-firing biomass with coal can never be a low-carbon fuel. Drax's most optimistic engineers concede that for technical reasons it can never exceed 20% biomass,[398] which means it will be at best 80% coal and therefore still far higher emitting than any other source of electricity apart from straightforward coal.

Large scale biomass power plants

A new 350MW biomass power plant, the largest in the world, has been granted permission in Port Talbot.[315] The plant £400 million plant[316] will require 2.5-3 million tonnes of 'carbon neutral' wood chip each year, which the company, Prenergy, plans to import from countries such as Russia and Ukraine.[317]

With large scale power plants such as this, and E.On's Stevens Croft plant in Scotland and Blackburn Meadows plant in Sheffield,[318] a significant portion of the energy is wasted as heat.

Sustainable use of biomass

So co-firing coal with biomass or burning biomass in existing large plants is no solution. But biomass does have a role to play where it can be harvested and used sustainably on a smaller scale, either providing heat to individual buildings or for combined heat and power. To decide whether or not a biomass project is sustainable, the full life-cycle emissions from the fuel including from land use change, transportation and processing must be taken into account. Recycling waste products, including waste wood, has the potential to qualify as sustainable biomass where no harmful emissions are involved (as there are with conventional waste incinerators). Impacts on soil depletion from using agricultural waste should also be considered.

Traditional biomass

Traditional use of biomass also has its own associated issues: the World Health Organisation estimates that two million people in the Majority World, predominantly women and young children, die prematurely each year from breathing the fumes from indoor biomass stoves.[319] In addition, efficiency of traditional biomass use (especially open fires) can be low and use of wood for fuel is a significant contributor to deforestation, especially in Africa. Users of traditional biomass need to be enabled to develop improved technologies.

Biogas

Biogas is usually produced from agricultural waste including manure, or from sewage. The material is broken down by the natural action of bacteria in a sealed container called a digester, allowing the gas to be piped off and burned for heating, cooking or power generation. The remaining solid waste can be used as fertiliser if it is not contaminated with other materials (use of sewage waste in this way can be hazardous).

Small biogas digesters using animal manure to supply gas to households or villages are already used in some parts of the Majority World, particularly India. Farm-scale biogas digestion plants are increasing in popularity in Germany and Scandinavia. Local use of materials makes most sense to avoid transporting bulky, smelly raw materials. While the technology could only supply a small portion of total energy needs, it has relatively few negative side effects and is suitable for use in areas without infrastructure such as electricity grids or piped natural gas.

Scores: Biomass

Effectiveness at reducing greenhouse gas emissions: 6/10 – potential reduced by transport and processing emissions
Living up to the hype (science to spin ratio): 1:3
Democratic ownership and control: 9/10 (biogas and traditional biomass use)
3/10 (electricity/heat generation from biomass) due to dominance by large energy corporations
Social justice: 7/10 (biogas)
2/10 (electricity/heat generation from biomass) food price rises, land-grabbing
Sustainability: 10/10 (biogas)
3/10 (electricity/heat generation from biomass) current or slightly expanded levels should be sustainable but further growth relies on land-use change
Scalability: 3/10 (overall)

Incineration & Landfill Gas

Two technologies sometimes classed under biomass use (but not assessed as such elsewhere in this section) are landfill gas and incineration. Neither of these makes a significant contribution to reducing greenhouse gas emissions from energy generation, but they are sometimes suggested as means of dealing with the emissions from waste.

Landfill waste dumps generate methane as degradable waste breaks down with insufficient oxygen. Methane is a greenhouse gas 25 times as potent as carbon dioxide[287]. Some of the methane can be captured (possibly up to 75%, more often 25-50%) and used as fuel. While this may reduce emissions from existing landfill sites, emissions per unit of energy remain higher than from a gas-fired power station. More importantly, other methods of waste disposal, especially separation of waste at source with materials recycled, composted or anaerobically digested to produce biogas, achieves much greater overall emissions reductions.[320]

Incineration of waste in the UK usually means burning the waste that remains after metal and plastics have been separated for recycling, and using the heat to generate electricity, as in a coal or gas-fired power station. Much of this 'residual' waste consists of plastics, derived from fossil fuels, which give off carbon dioxide, other greenhouse gases and other pollutants when burned. Electricity only incinerators emit 33% more carbon dioxide than gas fired power stations. They also rely on maintaining high levels of waste.

Decentralised Energy vs. Extending the Grid

How it Works

Decentralised energy

Generating energy through combustion of fossil fuels at centralised power stations and transmission through the National Grid is incredibly wasteful. 61.5% of the energy is lost as heat at the power stations, 3.5% is lost in transmission as the electricity has to travel through the grid at high voltage and then be converted to low voltages to be used domestically.[321]

The solution to this put forward by Greenpeace and the European Renewable Energy Council (amongst others) is to have local grids.[322] In this scenario local areas would produce electricity using a variety of renewable technologies. As well as small wind farms, solar panels and other medium-scale technologies, homes and businesses would 'micro-generate' their own energy. This would be backed up by 'combined heat and power' (CHP) plants; small power stations in urban areas burning fossil fuels or biomass which distribute their surplus heat to nearby buildings via insulated hot water pipes.[323] This gives much more efficient overall use of energy since the heat produced generating electricity is not wasted.

Extending the grid

On the other end of the scale, there is the suggestion of extending electricity grids outwards, connecting up the UK grid with Europe and beyond to harness the enormous potential for concentrated solar power from the Sahara desert. The plan is called TREC - Trans-Mediterranean Renewable Energy Co-operation[324] – and involves linking up all renewable energy resources and trading them across Europe, North Africa and the Middle East with new efficient high voltage DC cables.[325]

Decentralised Energy and Extended Grid Current Position

Status of the technology

Decentralised energy mainly uses existing technologies which have been successfully in use for decades, particularly in other parts of northern Europe such as Denmark. For more on the micro-generation component, see the relevant sections above.

Extending the grid would rely on use of high voltage DC cables, for which the technology exists but is not in widespread use.

Major players

In the UK, decentralised energy is being promoted by various local authorities, including London and Edinburgh, and NGOs including Greenpeace and World Wide Fund for Nature (WWF).

Extending the grid is supported by TREC, Trans-Mediterranean Renewable Energy Co-operation. TREC is a project of the founded in 2003 by The Club of Rome, the Hamburg Climate Protection Foundation and the National Energy Research Center of Jordan (NERC).

Decentralised Energy and Extended Grid Issues

Issues with decentralised energy

Decentralised energy has the advantages that it gives communities more autonomy over their energy supply and it is more efficient. Also, using combined heat and power gets over the problem of how to heat homes effectively without high carbon dioxide emissions. Most plans still require some connection to a national grid, as a minimum for backup purposes.

The idea has its disadvantages though. Micro generation of electricity through mini wind turbines is much less efficient than large scale projects. District heating networks would require installation of expensive new infrastructure which would generally only be viable in densely populated areas. CHP plants need a sustainable fuel source. CHP would offer a more sustainable way of using remaining gas reserves – but is not a long term solution, particularly with the UK relying increasingly on imports of liquefied natural gas which has a high carbon cost from liquefaction and transportation. CHP means lower emissions but it's a long-term commitment to fossil fuel use with no possibility of carbon capture and storage.

Issues with an extended grid

An extended grid would help to eliminate problems of intermittency of supply as there should always be wind blowing or sun shining somewhere in the extended region. It is predicted that 10-25% of Europe's energy could come from concentrated solar thermal energy from the Sahara by 2050.[326] The question this raises is whether the richer nations that are part of the programme would end up effectively colonising the renewable energy resources of the poorer nations, as has happened in the past with other natural resources from oil to gold to timber, contributing to poverty and conflict in Africa and elsewhere. 80% of people in Sub Saharan Africa have no access to electricity.[327] Sub-Saharan Africa is not currently part of the TREC programme.

Don't these countries also have rights to share in Africa's electricity generating potential? At present, since they don't have the money to invest in the project, they don't get the power.

Conclusion

Decentralised energy and an extended grid are not mutually exclusive. Both may be required to maximise use of renewables, alongside an intelligently organised, equitable supply and demand management system.

Scores: Extended Grid

Effectiveness at reducing greenhouse gas emissions: 6/10 - could encourage uptake of renewable technologies
Living up to the hype (science to spin ratio): 1:4
Democratic ownership and control: 2/10 - takes control of electricity generation further away from individuals and communities
Social justice: 3/10 - possibility of 'energy colonialism'
Sustainability: 8/10 - once established, system should be robust
Scalability: 6/10 - not proven what proportion of needs could be met

Scores: Decentralised Energy

Effectiveness at reducing greenhouse gas emissions: 6/10 - more efficient use of existing fuel, but doesn't eliminate fossil fuel and microgeneration may be inefficient
Living up to the hype (science to spin ratio): 1:2
Democratic ownership and control: 7/10 - transfers power to local government and smaller power companies
Social justice: 7/10
Sustainability: 5/10 - lowered due to most DE systems currently relying on gas-fired Combined Heat & Power (CHP)
Scalability: 6/10 - could be expanded considerably, but unclear what proportion of UK needs would be met

CHAPTER V: THE TECHNOLOGIES - GEOENGINEERING

Things must be getting really bad when people suggest terraforming[328] our own planet. Not only that but these people are being taken seriously. The term geoengineering refers to the large scale manipulation of the environment to bring about specific environmental change, particularly to counteract the undesirable side effects of other human activities.[329] The geoengineering community prefers to refer to its work as 'Earth Systems Engineering and Management'[330] to avoid the dodgy sci-fi overtones. Technologies proposed include blasting the stratosphere with sulphates, mirrors in space, covering the deserts in reflective plastic, and dumping iron fertiliser in the oceans. Geoengineering technologies are presented as 'last ditch' options. But at bottom geoengineering is all about sounding just plausible enough to perpetuate the myth that there is an easy way out of this crisis. And about money from carbon credits, of course.

How the Technologies Work

There are various geoengineering projects on the table and they broadly fit into two categories: technologies to increase the amount of the sun's energy that is reflected back into space, and technologies to remove carbon dioxide from the atmosphere. These methods differ from carbon capture and storage in that they remove carbon dioxide once it has been dispersed in the atmosphere rather than at the point of emission. This requires planetary-scale engineering of ecosystems to have any significant impact on atmospheric carbon dioxide levels.

Geoengineering Current Position

Status of the technology

Most of the technologies suggested are purely theoretical. A few, including ocean fertilisation and capture of carbon dioxide from the air, are beginning to be commercialised, though they remain controversial and highly speculative, with major uncertainties surrounding their effectiveness.

Support for the technology

Geoengineering is seen by many scientists as an extreme and potentially dangerous approach to climate change mitigation. However there has been government interest. In 2001 a conference was held at the White House between the President's Climate Change Technology Programme and scientists researching geoengineering, entitled 'Response Options to Rapid or Severe Climate Change'.[331] The USA also pushed for geoengineering to be included in the Intergovernmental Panel on Climate Change's report into the potential for climate mitigation.[332] In 2006, NASA held a closed meeting to discuss geoengineering including the use of sulphates.[333] The USA was also the key nation at the last meeting of the London Convention (the International Maritime Organization body that oversees dumping of wastes at sea) pushing for a watered down statement concerning the legality and impact of ocean fertilisation.[334]

In the UK, an unnamed 'senior environment minister' commented that: 'The data on ocean acidification over the past year has turned heads in the upper levels...ocean fertilisation, because of its enormous potential simply must (I will emphasise the word must) be explored vigorously... Other governments in Europe and Asia are of the same opinion and are moving forward with their own plans. The question is how to do this without engendering public opposition.'[335]

Ocean fertilisation experiments have been funded by countries including Australia, Canada, Japan, Germany, Mexico, Netherlands and New Zealand.[336]

Major players

Ocean fertilisation companies:
Climos, Atmocean, Planktos [now defunct], Ocean Nourishment Corporation

Carbon sequestration:
Synthetic Genomics, Global Research Technologies

Carbon offsets:
Carbon Neutral Company, Climate Care, JPMorgan, Atmosfair, Natsource, Terrapass, DrivingGreen, Drive Neutral, carbonfund.org, MyClimate, AtmosClear Climate Club, Carbonfund.org, Climate Friendly

Investors:
Virgin Earth Challenge, Breamar Energy, Elon Musk

Academics – supportive of or researching geoengineering to varying degrees:
Paul Crutzen (sulphate aerosols), Ken Caldiera (albedo enhancement), Roger Angel (mirrors in space), Stephen Slater (cloud seeding), Alvia Gaskill (albedo enhancement), David Keith, John Latham (cloud seeding), Klaus Lackner (carbon sequestration).

The Technologies

Albedo enhancement - Reflecting the sun's energy

The capacity of the earth to reflect back solar energy is know as its 'albedo'. Methods to increase this reflective capacity are also known as 'albedo enhancement'. Some of these technologies sound like a joke – but they're not.

Sulphates in the stratosphere

When volcanoes erupt they release sulphates which are known to have a cooling effect on global temperatures by reflecting solar energy back into space. The eruption of Mount Tambora in 1815 led to the 'year without summer'.[337] Planet-wide episodes of cooling also followed the eruptions of El Chichón (1982) and Mt. Pinatubo (1991).[338] Some scientists are therefore proposing to increase the level of sulphates in the atmosphere to simulate the effect of a volcanic eruption. Methods for getting the sulphur up to the stratosphere include using tens of thousands of balloons,[339] using naval cannons,[340] big pipes or specially designed aircraft[341]. Another suggestion is to increase the sulphate emissions of conventional aircraft by using fuel with a higher sulphur content.[342] It is estimated that five million tonnes of sulphur would be needed each year to compensate for a doubling of carbon dioxide concentrations. Costs are estimated at around $25-50bn (£12.5-25bn) a year.[343]

Beyond the overarching problems with albedo enhancement as a solution to climate change, there are a number of significant issues with this particular approach. It is essentially fighting pollution with more pollution. Sulphate pollution causes a thinning of the earth's ozone layer. The sulphates will eventually come back down to earth, with an unknown impact on ecosystems. Governments have been working to reduce emissions of sulphates because they cause acid rain. Nobel prize winner Paul Crutzen, who advocated research into sulphate aerosols as a last ditch solution to global warming, predicted around half a million deaths as a result of particulate pollution.[344] New studies have shown that the historic droughts in the Sahel region of Africa that caused widespread famine in the 1970s and 80s were caused in part by industrial emissions of sulphates in the West.[345] The clean air acts that reduced this pollution helped trigger the return of the rain to the Sahel. It is impossible to predict which regions would be affected by this climate manipulation or how.

Plastic coated deserts

According to Alvia Gaskill of the Global Albedo Enhancement Project, covering the Sahara, Arabian and Gobi deserts with polythene sheeting could delay the effects of global warming by 60 years.[346] In this plan 67,000 square miles of desert would be given the shiny plastic treatment each year for 60 years.[347] The plastic sheeting would have to be maintained, and periodically replaced, for a century or two[348] (which could be a logistical issue in desert regions). Premature removal would have a rapid global warming effect.

Issues with this particular approach include the possibility of significant and unpredictable changes to regional weather patterns and changes to the local, regional and even global climate. This could particularly exacerbate poverty in

Africa. All life in the affected desert region would be eliminated. Also, dust from the Sahara travels thousands of miles and transports nutrients way out into the Atlantic Ocean, forming part of the basis of the ocean's food chain.[349]

Gaskill fails to address the issue that plastic is made from oil, oil supplies are running low, and plastic manufacture has its own greenhouse gas emissions. On the upside, Alvia Gaskill suggests that this method would provide valuable experience in terraforming which could be used in future interplanetary space exploration, and lower global temperatures would lessen demand for air conditioning and so reduce carbon emissions.[350] Satire is dead.

Sunshades in space

Another albedo-enhancement scheme involves a set of 16 trillion transparent, sunlight-refracting shades about 1.5 million km from Earth. Each shade would weigh only about a gram and cover roughly the area of a broadsheet newspaper page. The project would require 20 launchers each positioning 800,000 screens every five minutes for ten years and would cost trillions of dollars over 25 years.[351] This scheme would not only involve high cost but also massive amounts of energy, and the technology is, to put it mildly, highly speculative.

Making clouds

Clouds are white so they reflect sunlight. Increasing the amount of clouds will reflect back more solar energy. But increasing cloud cover affects the weather so to avoid significant impact on humans the proposal is to increase cloud cover way out in the open ocean. The method proposed is to use specially designed boats which whoosh up ocean water into the air. The idea's exponents claim that spraying approximately 50 cubic metres of ocean water per second would counteract a doubling of present-day concentrations of carbon dioxide.[352]

Shiny plants

Plant shinier varieties of agricultural crops.[353] Use would have to be widespread to have any significant albedo enhancement effect. As with agrofuels, this could potentially have an impact on food prices and deforestation rates due to land use change and increase the use of genetically modified crops.

Paint everything white

Yes, everything that can be painted[354] – roofs, roads, you name it. It doesn't seem to have occurred to the proposer that driving on white roads on a sunny day would cause crashes everywhere as drivers are dazzled by the light. And if the roads aren't for driving on then dig them up, don't paint them.

Cover the sea with ping pong balls

To mimic the reflective effect of arctic ice, thereby reducing the amount of heat absorbed by the earth's surface. Okay, so that was a spoof[355] – but could you tell the difference?

Overarching Issues with 'Albedo Enhancement'

Once a scheme of albedo enhancement has been embarked upon it must be maintained for as long as the carbon dioxide emissions that it aimed to counteract remain in the atmosphere (up to 200 years) regardless of any negative impact the scheme is found to have. Humanity is already conducting an uncontrolled planetary scale experiment with the planet's climate through greenhouse gas emissions. Is it really sensible to start another one which could have equally disastrous and unpredictable consequences? Especially when other more certain and less damaging options are already available ?

Albedo enhancement also does nothing to counteract the other effects of high carbon dioxide concentrations in the biosphere. Dissolved carbon dioxide is making the oceans more acidic.[356] If carbon dioxide emissions continue, the pH of the oceans is predicted this century to reach a level lower than for the past hundreds of millennia. Ocean acidification is already destroying the world's coral reefs and also threatens organisms such as phytoplankton and zooplankton which are the basis of major food chains. It is unclear whether marine species and ecosystems will be able to adapt quickly enough to survive.[357]

Removal of Carbon Dioxide from the Atmosphere

In February 2007, Richard Branson announced the launch of the Virgin Earth Challenge, a prize of $25 million for the person who can 'demonstrate to the judges' satisfaction a commercially viable design which results in the removal of anthropogenic, atmospheric greenhouse gases so as to contribute materially to the stability of Earth's climate.'[358] Judges include Branson, the former US Vice-President and chief negotiator responsible for wrecking the Kyoto Protocol, Al Gore, and maverick ecologist James Lovelock.[359]

Removal of carbon dioxide from the atmosphere poses difficulties because of the low concentrations of carbon dioxide in the air. Approximately 0.04% of the air is made up of carbon dioxide. This is enough to have a significant warming effect, but it makes capture very difficult.

Ocean fertilisation

One set of schemes for carbon dioxide capture centres on encouraging the growth of phytoplankton in the oceans, which take up carbon dioxide as they photosynthesise. In theory, some of this carbon dioxide might not return immediately to the carbon cycle as some of the plankton's carboniferous bodies fall to the ocean floor or are eaten by other marine organisms, which will excrete carbon pellets which may drop to the ocean floor. Exactly how much carbon dioxide is sequestered by this method, and for how long, has not been quantified.

A number of companies are already trying to commercialise ocean fertilisation, hoping to make a great deal of money from carbon credits. To make plankton grow the companies are proposing various methods of artificially adding nutrients to the oceans. Methods include dumping tiny iron particles,[360] ammonia or urea,[361] or piping up nutrient-rich deep ocean waters to the surface.[362]

Ocean scientists, including the IPCC,[363] have warned that this technology is potentially dangerous to ocean ecosystems, unlikely to sequester much carbon dioxide, has the potential to increase levels of other dangerous greenhouse gases such as nitrous oxide and methane, increase ocean acidification in deep ocean waters, deplete nutrient loading in surface waters potentially leading to 'dead zones', and other unpredictable effects.[364] The lifecycle emissions involved in producing and dispersing the fertiliser is also an issue.

In a recent *Science* article, key scientists (including some that had previously been positive about ocean fertilisation) highlighted concerns, concluding that 'it is premature to sell carbon offsets from the first generation of commercial-scale OIF [ocean iron fertilisation] experiments unless there is better demonstration that OIF effectively removes carbon dioxide, retains that carbon in the ocean for a quantifiable amount of time, and has acceptable and predictable environmental impacts'.[365] There has been a significant civil society criticism of plans to proceed with ocean fertilisation. One company, Planktos, which had been planning to sell carbon credits for ocean fertilisation, announced that it was folding, blaming a 'highly effective disinformation campaign waged by anti-offset crusaders'.[366]

Despite opposition, scientific uncertainty and potentially huge environmental risk, a group of companies - Climos, Ocean Nourishment Corporation and Atmocean - are proceeding with their schemes, exploiting the fact until recently there has been no regulatory oversight over this type of activity. In November 2007 the London Convention (the International Maritime Organization body that oversees dumping of wastes at sea) spoke out on the issue of ocean fertilisation, stating that it is not justified and requires oversight and regulation, and that given current evidence large scale ocean fertilising schemes would be in contravention of the Convention.[367] This is a big step forward. Although the Convention only applies to the 80 signatory states, organisations approving carbon credits are much less likely to grant credits for a scheme that is deemed to breach an international Convention.

The companies haven't given up that easily, however. In December 2007, Climos announced the first steps towards carbon credits for ocean fertilisation were under way as a draft methodology had been submitted to auditors Det Norske Veritas and has appointed California based Tetra Tech to do a environmental impact assessment.[368] The company has also raised $3.5m (£1.8m) venture capital funding,[369] and is planning to spend the money undertaking research dumps in 100km2 areas of ocean which they expect will be underway in 2009.[370] The company plans to co-opt independent scientists who have previously been involved in ocean fertilisation experiments to conduct its research.[371] Climos claims to be a responsible ocean fertilisation company and has proposed a voluntary code of conduct for commercial ocean fertilisation.[372] The company's presentation of its voluntary code at the London Convention meeting did not, however, appear to convince the delegates.[373] Since the impacts of ocean fertilisation are so little understood a code of conduct which allows potentially dangerous activity to go ahead is meaningless.

On the other side of the world, the Ocean Nourishment Corporation has approached the Philippine government for permission for a dump of urea in the Sulu Sea.[374] The urea is synthetic, made from natural gas, which shows how seriously they are taking the issue of sustainability. Ocean Nourishment Corporation is expected to claim fishing rights on the basis of increasing fish stocks because of the phytoplankton blooms. This has outraged Philippine fishing communities. Ruperto Aleroza of Kilusang Mangingisda, the Philippine Fisherfolks Movement, said 'This technology is unacceptable. It is a dangerous technology that could imperil the marine environment which is the main source of survival and livelihood of poor fisherfolk in the Philippines. Under Philippine laws experiments like this must undergo environmental impact assessments and free prior informed consent of communities that are potentially affected.' [397]

While it is good to see that regulation is coming in for ocean fertilisation, this only serves to highlight the fact that there is no forum for making decisions about this kind of technology. When it comes to engineering the planet's natural systems, one country, or company, may make a decision which has huge implications for the rest of the planet, but there is currently no decision-making body that can stop them or hold them accountable.

Planting trees

Planting trees to offset carbon emissions has been widely discredited by environmentalists[375] but the sale of these 'carbon offsets' has by no means disappeared. The theory is that through photosynthesis trees absorb carbon from the atmosphere and lock it away within the cell structure of the wood. However, this is in no way a reversal of emissions from fossil fuels. Fossil fuels introduce carbon dioxide into the atmosphere which has been stored for millions of years. Trees are part of the biological carbon cycle – the interchange of carbon in different forms between the atmosphere, living things and the oceans. Even the carbon temporarily sequestered in treetrunks is usually released again within decades when the tree rots or burns.

A further issue is the time lag. A tree absorbs carbon slowly over many decades. But the carbon dioxide emissions that the tree is supposed to offset take place in a short time frame. In the period between emission and absorption the emissions are already having a damaging climate impact. This is like pouring water into a boat by the bucket load and claiming it won't sink as you are also removing water the with a teaspoon. In addition to this, offset schemes sell the offsets long before the offsetting project begins so again the carbon dioxide already in the atmosphere is contributing to global warming in that time.

We can expect the sale of offsets to grow, however, as one of the leaders in the carbon offset industry in the UK, Climate Care, has recently been bought out by financial services giant JP Morgan.[376]

Burying trees

As an answer to the problem that carbon in trees remains part of the carbon cycle, periodically returning to the atmosphere, it has been suggested that trees should be buried, in an attempt to keep their stored carbon out of circulation for longer.

The proposal on the table from one US scientist is to take dead wood from the forest floor and selectively logged old trees from the world's forests and bury them in trenches.[377] The paper estimates that the cost of this method of carbon sequestration would be $14 (£7) per tonne of carbon dioxide and that if all the world's forests were managed in this way there would be the theoretical potential to sequester five gigatonnes of carbon dioxide a year, equivalent to about three times annual emissions from land-use change during the 1990s.[378]

However, this is a purely theoretical measure. Consider, for example, the emissions from the diggers digging all those trenches. Consider the massive damage to natural forests from the removal of dead wood, which forms a vital part of ecosystems, and from the burial trenches and all the roads necessary to collect wood and access trenches – roads which would also open up the forests to illegal loggers and poachers. Consider the sheer difficulty of accessing most old-growth forest – there are vast areas where such an intervention would never be practical. In any case, even buried wood rots eventually, releasing its stored carbon.

Growing trees in new forests specifically for carbon sequestration purposes would encourage the use of genetic modification of trees to increase their absorption of carbon dioxide and encourage monoculture plantations (with all their environmental impacts – see section on agrofuels). Putting a carbon price on dead wood would also impact on the price of wood, and so affect logging activities, deforestation and collection of wood for fuel, which would be impossible to accurately predict. A scenario could be envisaged where loggers clearing rainforest for plantations are awarded carbon credits for burying rather than burning the logged trees.

Biochar and Bioenergy with Carbon Storage (BECS)

Two proposals seek to address the issue of removing carbon dioxide from the air at the same time as generating low carbon energy: biochar and bioenergy with carbon storage (BECS). In BECS, biomass is used to generate heat and power together with carbon capture and storage, this is argued to be carbon negative as the plants absorb carbon dioxide from the air as they grow, and this is captured and stored underground removing it from the carbon pool. The BECS proposal seeks to avoid the onset of abrupt climate change tipping points by quickly converting a large amount of land to biomass.[379] One figure used is 500 million hectares globally by 2030[380] an area larger than the entire European Union. Proponents argue that, in combination with widespread use of energy efficiency and non fuel renewables such as wind and solar, BECS could sequester enough carbon to restore concentrations of carbon dioxide to pre-industrial levels.[381] Carbon capture of biomass would, however, be dependent on the infrastructure existing to transport and store carbon dioxide. This would limit this form of biomass use to large centralised power stations, making efficient use of the heat produced difficult. Also, it wouldn't have large-scale impact until grown, harvested and the carbon captured and stored. As we've seen, that's decades away. This isn't going to be at all possible within the necessary timeframe.

A similar approach is taken with biochar, where biomass is burnt without oxygen (a process known as pyrolysis) creating biogas, bio-oil, and biochar.[382] Biogas and bio-oil can be used to generate heat and power, while the carbon rich biochar is returned to the soil where experiments suggest carbon can be sequestered for hundreds of years.[383] The proportion of carbon dioxide sequestered is variable but a figure widely used is 20%. One study suggests that it would be possible to 'offset' 10% of the annual US fossil-fuel emissions using either 'forest residues' from the 200 million hectares of US forests that are used for timber production, by growing fast growing biomass on 30 million hectares of conservation reserve land, or collecting crop residues from 120 million hectares of cropland.[384] In addition to sequestering carbon, biochar is also said to improve soil structure, and the retention of water and nutrients, reducing leaching of nitrates into water sources and emissions of nitrous oxide (a potent greenhouse gas). However, co-producing biochar is less efficient than other uses for bioenergy, there has been limited research into potential applications for bio-oil, and biochar production is reliant on supplies of cheap biomass.[385]

Proponents have characterised these proposals as 'benign low risk geoengineering'.[386] However, the proposals hit the same problems as the other types of large scale bioenergy use already discussed in this report. Bioenergy takes energy inputs to grow, process and transport, and the need for cheap feedstocks requires monoculture intensive farming methods or the removal of 'agricultural waste' from the soil, advancing soil depletion. More critically, land is a finite resource. Since human land use for settlement and agriculture is already placing severe stress on remaining natural ecosystems, any major expansion of bioenergy has massive indirect carbon emissions which appear not to be factored into these proposals.

Synthetic trees

A company called Global Research Technologies is seeking to commercialise artificial trees. The 'trees' - towers shaped like giant fly swatters[387] - use chemicals[388] to absorb carbon dioxide. The intention is that the carbon dioxide is then pumped down oil wells as with carbon capture and storage from coal power plants – with the same implication that it could be used for enhanced oil recovery. The company hopes to scale up its prototype which currently is able to absorb five tonnes of carbon dioxide per year, and claims that it will be able to install 100 devices over the next five years, each absorbing one tonne of carbon dioxide per day.[389]

However, the design uses a significant amount of energy. Energy is needed for the manufacture of the 'trees', for the production of the chemicals in the 'trees', for the transport and storage of the carbon dioxide, and, most importantly for removing the carbon dioxide from the chemical solvent so that it can be used to capture more carbon dioxide.[390] The prototype 'tree' requires 'intensive use of electricity' to do this, in a process called electrodialysis. It is unclear from the company's information what effect this has on net emissions or savings of carbon dioxide if the electricity used is fossil fuel generated. The company claims that the energy used for capturing the carbon dioxide is 'comparable to or less than that of carbon dioxide capture in the flue stack of a conventional power plant'.[391] But also that 'if this electricity were produced by a conventional coal-based power plant, the carbon dioxide capture would exceed the carbon dioxide release. Net capture is positive but small.'[392]

Other methods for capturing carbon dioxide from the air have also been proposed including by synthetic biology company Synthetic Genomics, led by Craig Venter, who claims that he will be able to develop a synthetic microbe that will be able to convert carbon dioxide into bioenergy.[393]

Overarching Issues with Geoengineering

Geoengineering rests on the assumption that humans are masters of the universe and the natural world and can control and engineer its systems – that humans can run ecosystems better than nature can. The overwhelming evidence to the contrary drawn from the current state of ecosystems humans have manipulated is ignored.

Further, the geoengineering approach starts from the assumptions that the behaviour of societies cannot change and that energy demand growth is essential and inevitable. If one believes it is impossible to do anything about emissions levels, then maybe covering the earth with synthetic trees, dumping fertiliser in the oceans and blasting the stratosphere with sulphur at the expense of the delicate balance of many of the earth's vital ecosystems are appealing ideas, even if they are highly risky, completely experimental and potentially ecologically catastrophic.
Putting these ideas into practice would be dangerous and would not work. Climate change has shown that humans do not and probably never will understand the planet's systems well enough to try to artificially engineer a re-balancing of the scales that overconsumption has tipped. Rather than press an ill-judged thumb on the other side of the scales, humanity should be reducing its impacts, restoring natural ecosystems and letting the planet find a new equilibrium.

The implications of geoengineering technologies at a planetary scale are unknown. There would be no way back after embarking on such a course of action.

Carbon credits

Many suggested geoengineering schemes simply offer a get-out clause to big emitters and potentially lucrative returns to those that develop and invest in the technologies. Ocean fertilisation projects and tree plantations are little more than a scam to make money on the carbon market. Anyone who thinks that ocean fertilisation would be used as a last resort to help society smooth a period of extreme adaptation is in denial about the realities of the carbon market and the intentions of ocean fertilisation companies.

Even if one ignores the potential side effects of pursuing geoengineering schemes, it has to be taken into account that they will not be used help achieve increased cuts in carbon emissions, as any emissions cuts these schemes might make would be traded away to whichever large polluter wants the credits. This is why venture capitalists are interested. If geoengineering were being pursued only as a last resort option linked with strong emissions cuts then no venture capitalist funder would touch it because there would be no hope for them to gain significant financial returns.

Unilateral action

Even if geoengineering technologies are widely rejected the possibility remains of an individual country acting unilaterally and going ahead with a large scale geoengineering project that would have consequences for the whole world. This problem could only be addressed by pre-emptive international co-operation.

Conclusion

As with many other techno-fixes, geoengineering promises technological breakthroughs at an unspecified point in the future, and so helps to assure people that there is no need for major change to the way society works or the rate at which people consume energy and products. As journalist Johann Hari put it: 'carrying on pumping out greenhouse gases because of the possibility of geo-engineering is like telling an alcoholic that he doesn't need to quit drinking, because in a few years you'll give him a liver transplant – with a few rusty old knives you found in your garage.'[394]

Scores: The Technologies

Effectiveness at reducing greenhouse gas emissions: 2/10
Living up to the hype (**science to spin ratio**): 1:10
Democratic ownership and control: 1/10 proposed schemes all 'top-down'
Social justice: 1/10 negative effects on (for example) desert-dwellers, forest-dwellers, fishing industry, victims of artificially-induced weather patterns, future generations required to maintain systems.
Sustainability: 1/10 – schemes would need continuing over long periods with large inputs of effort
Scalability: 1/10 because no evidence of emissions reductions to scale up

CHAPTER VI: CONCLUSION

> ***Continued rapid growth of CO2 emissions and infrastructure for another decade may make... [avoiding dangerous climate change]... impractical if not impossible.[395]***

Politicians and corporations aren't just getting our climate change solutions wrong because they're looking at the wrong technologies, they're only looking at this one facet of the change that is needed. As we outlined in chapter 2, we need change in many areas: technological change, political change, economic change and social change. They are all interlinked, but social change drives them all.

This report has not presented a manifesto. We should mistrust anyone who thinks they have all the answers – it is never that simple. This is a contribution to a conversation which we are all engaged in right now and which demands to be everyone's main focus at this time of crisis.

Technological change

Technological change is not sufficient in itself but it is vital. We need to have criteria for sustainable technologies. Many technologies which are conventionally termed 'renewable' (agrofuels and other agro-based bioenergy, large scale hydro, landfill gas and waste incineration) contribute to the problem. Technologies that are available now, and help to engender a wider shift towards sustainable societies are the ones which we should focus on.

Research and development is needed to increase the efficiency of these technologies, and to improve energy storage and demand management technologies which support sustainable renewable energy production.

Political change

Climate mitigation is a political and not a simply economic issue. Decisions need to be taken as a society as to the solutions which we are going to adopt. This requires democratic engagement and government action through policy making, not to create market incentives but to actively promote changes to the way energy is produced and consumed throughout society.

Economic change

As the 'consumption era' comes to its inevitable decline, we need to find ways of managing a just transition to a society whose economy is based on meeting needs sustainably and equitably.

The current economic model is not up to the job. Quantifying the economic 'value' of acting or not acting in the face of climate change is futile since critical impacts (loss of life, irreversible ecosystem collapse) cannot practically or morally be given a simple numerical value. The impetus to act must be driven by our scientific understanding of the problem, our decisions as to how to act must be driven by our assessment of which measures are most effective, just and sustainable rather than simply which are most economically attractive.

Social change

Social change is what will cause all of the rest of the changes to happen if they are to happen. It is what has driven the changes that have happened so far.

The social change approach means taking systems that are unsustainable and finding ways of meeting people's needs within the limits of the planet's resources through co-operation, lifestyle change and appropriate technology. We need to find sustainable ways of managing our food systems, transport systems, housing, land use and economic activity. These sustainable solutions will, for the large part, be small scale and localised, with solutions meeting the needs of local populations. Achieving this means co-operation at a community level.

Some people argue that if climate change is a global problem, the solutions be implemented on a global, or at least grand, scale? But this ignores a lot of complexities: large scale solutions do not take into account the small scale and localised impacts and the distributed nature of the causes of climate change. Large scale solutions are more measurable. Replacing one fuel source with another can be shown to reduce a quantifiable amount of carbon emissions. Whereas emissions reductions through local level interventions or a change in behaviour patterns can be much more difficult to predict or measure. Social changes are less easy to understand when approaching the problem of greenhouse gas emissions for the point of view of a systems analyst or carbon accountant. Which is why the mechanisms for reducing emissions, such as carbon trading, fails to take them into account.

Frequently when people working for social change put forward the kinds of changes that are needed they are told that they are 'living in fairyland'. Not that they are wrong, just that they are unrealistic.

You want fairyland? Try this:

- Continue emitting greenhouse gases, making the problem massively worse, for another twenty years, in the hope that something will come up.
- Rely on a set of technologies which haven't even been built yet, and in some cases will never work.
- Do nothing about the third of emissions which come from deforestation and agriculture.
- Do nothing about the underlying cause of the problem – overconsumption of natural resources.

You may not end up living in fairyland, but you'll find yourself on an unrecognisable planet. And the plastic-coated deserts, fake trees and cloud-making machines have a nice fantasy feel to them.

So let's get realistic. Technologies are a useful part of the solution, but techno-fixation isn't. Other changes are even more important than technology, and equally technically possible. Whether or not they are achieved depends on the actions we take now. The obstacles that stand in the way of solutions to climate change are great. But with all the knowledge and experience we have, and all our resourcefulness and adaptability to change, for as long as we haven't crossed the two degrees threshold we have a chance. And not taking that chance is suicide.

The beginning of the path towards a sustainable solution to climate change could look something like where we are now. The science is uncontroversial. There is a groundswell of public opinion. Politicians and corporations are giving lip service to the solutions but the public do not trust them. Mainstream politicians are even starting to question the logic of perpetual economic growth. The seeds for change are being sown. There's a huge amount of work to be done to make these seeds germinate and flourish, but it can be done. There is still time.

References

1. An exhaustive treatment of the impacts of climate change can be found in the Intergovernmental Panel on Climate Change, Assessment Report 4, Working Group II. http://www.ipcc.ch

2. World Health Organization, 'Climate and Health Fact Sheet', July 2005 http://www.who.int/globalchange/news/fsclimandhealth/en/index.html

3. Intergovernmental Panel on Climate Change (IPCC), *Climate Change 2007: Synthesis Report: Summary for Policymakers*, 'Table SPM1', A1F Scenario, 2007, p8

4. *Avoiding Dangerous Climate Change, International Symposium on the Stabilisation of Greenhouse Gas Concentration*, Hadley Centre, Met Office, Exeter, UK, 1-3 February 2005, *Report of the International Scientific Steering Committee*, May 2005, p6

5. IPCC, *Fourth Assessment Report. Climate Change 2007: Synthesis Report. Summary for Policymakers*, 2007, Table SPM.6 www.ipcc.ch/pdf/assessment-report/ar4/syr/ar4_syr_spm.pdf

6. Joseph Canadell et al, 'Contributions to accelerating atmospheric CO2 growth from economic activity, carbon intensity, and efficiency of natural sinks' *Proceedings of the National Academy of Sciences*. 25 October 2007 www.pnas.org_cgi_doi_10.1073_pnas.0702737104

7. IPCC, *Climate Change 2007: Synthesis Report: Summary for Policymakers*, 'Table SPM1', A1F Scenario, 2007, p10-13

8. *Avoiding Dangerous Climate Change, International Symposium on the Stabilisation of Greenhouse Gas Concentration*, Hadley Centre, Met Office, Exeter, UK, 1-3 February 2005, *Report of the International Scientific Steering Committee*, May 2005, May 2005, p7

9. Lux Research, *The Cleantech Report*, 2007

10. IPCC, *Climate Change 2007: Mitigation of Climate Change. Contribution of Working Group III to the Fourth Assessment Report of the Intergovernmental Panel on Climate Change*, Cambridge University Press, 2007

11. IPCC, *Fourth Assessment Report. Climate Change 2007: Synthesis Report. Summary for Policymakers*, Table SPM.6, 2007. www.ipcc.ch/pdf/assessment-report/ar4/syr/ar4_syr_spm.pdf

12. Colin Forrest, 'Global and UK emissions reductions targets for 2030' *The Cutting Edge: climate science to April 2005.* http://portal.campaigncc.org/files/the_cutting_edge_climate_science_to_april_05.pdf

13. 'Blair bid for backing on climate', *BBC News*, 26 January 2005, http://news.bbc.co.uk/1/hi/sci/tech/4210503.stm viewed 2/2/08

14. JA Edmonds, MA Wise, JJ Dooley, SH Kim, SJ Smith, PJ Runci, LE Clarke, EL Malone, GM Stokes, *Global Energy Technology Strategy: Addressing Climate Change, Phase 2 findings from an international public-private sponsored research program*, May 2007, Battelle Memorial Institute

15. President George Bush, 'State of the Union Address', 23/01/07. Full transcript available at http://www.america.gov/st/texttrans-english/2007/January/20070123210844abretnuh0.9462549.html viewed 2/2/08

16. 'France gets nuclear fusion plant', *BBC News,* 29 June 2005, http://news.bbc.co.uk/1/hi/sci/tech/4629239.stm viewed 2/2/08

17. Johnnie L. Roberts, 'The Street Turns Green', *Newsweek*, March 2007 http://www.newsweek.com/id/36497 viewed 2/2/08

18. For a full analysis see George Monbiot, 'The Denial Industry' in *Heat: How to Stop the Planet Burning*, Penguin, 2006 p20-42

19. For a full analysis see Source Watch, 'Global Climate Coalition', http://www.sourcewatch.org/index.php?title=Global_Climate_Coalition viewed 2/2/08

20. For a full run down of court cases see: Gene Borio, 'Tobacco Timeline', http://www.tobacco.org/resources/history/Tobacco_History20-2.html viewed 2/2/08

21. High profile defections included BP, American Electric Power, Dow, Dupont, Royal Dutch Shell, Ford, Daimler Chrysler, Southern Company, Texaco and General Motors. See Source Watch, 'Global Climate Coalition', http://www.sourcewatch.org/index.php?title=Global_Climate_Coalition viewed 2/2/08

22. See: Larry Lohman, 'Made in the USA' in *Carbon Trading:A Critical Conversation on Climate Change, Privatisation and Power*, Dag Hammarskjold Foundation, October 2006, and George Monbiot, 'Hurray We're Going Backwards', The Guardian, 17 December 2007

23. See: Rebecca Spencer, *Corporate Law and Structures: Exposing the Root of the Problem*, Corporate Watch, January 2004 and Claire Fauset, *What's Wrong With Corporate Social Responsibility?*, Corporate Watch, May 2006 http://www.corporatewatch.org

24. Unilever, 'Promoting sustainable biofuels' October 2007. http://www.unilever.com/ourvalues/environment-society/sus-dev-report/climate-change/renewable-energy-biofuels.asp

25. Larry Lohmann, *Carbon Trading: A critical conversation on climate change, privatisation and power*, Dag Hammarskjold Foundation, Development Dialogue no. 48, September 2006, p195

26. ibid., p108-9

27. Suvi Monni, 'Uncertainties in the Finnish Greenhouse Gas Emissions Inventory', *Environmental Science and Policy* 7, 2004, p87–98

28. Michael Obersteiner et al., 'Quantifying a Fully Verifi able Kyoto', *World Resource Review* 14, 2002, p542.

29. 'Agency Slashes Check Monitoring of Industrial Emissions', *ENDS Report* 360, January 2005. See also Fred Pearce, 'Kyoto's Promises are Nothing but Hot Air', *New Scientist* 2557, 24 June 2006, p10.

30. Larry Lohmann, *Carbon Trading: A critical conversation on climate change, privatisation and power*, Dag Hammarskjold Foundation, Development Dialogue no. 48, September 2006, p132

31. ibid., p147

32. ibid., p141

33. ibid., p176-177

34. For analysis see Larry Lohmann, *Carbon Trading: A Critical Conversation on Climate Change, Privatisation and Power*, Development Dialogue No 48, September 2006, p101 - 121

35. Carbon Market Europe, 24 June 2005, http:www.pointcarbon.com

36. Open Europe, 'The High Price of Hot Air: Why the EU Emissions Trading Scheme is an Environmental and Economic Failure', Executive Summary, London, 2006, p3. http://www.openeurope.org.uk/research/ets.pdf,

37. Per Lekander, 'Guest Comment', *Carbon Market Europe*, 12 May 2006, p3

38. Larry Lohmann, *Carbon Trading: A critical conversation on climate change, privatisation and power*, Dag Hammarskjold Foundation, Development Dialogue no. 48, September 2006, p78-79

39. ibid., p118

40. ibid., p172

41. ibid., p181

42. ibid., p190-191

43. *Energy Needs, Choices and Possibilities: Scenarios to 2050*, Global Business Environment, Shell International 2001, p48 http://www.cleanenergyfunds.org/CaseStudies/Shell_2050.pdf

44. See for an analysis of this: Stefan Gossling and Paul Peeters, 'It Does Not Harm the Environment! An Analysis of Industry Discourses on Tourism, Air Travel and the Environment', *Journal of Sustainable Tourism* Vol. 15, No. 4, 2007

45. Ron Pernick and Clint Wilder, *The Clean Tech Revolution: The Next Big Growth and Investment Opportunity*, Collins, 2007 cover

46. ibid., p2

47. ibid., p19

48. Lux Research, *The Cleantech Report*, 2007

49. Ron Pernick and Clint Wilder, *The Clean Tech Revolution: The Next Big Growth and Investment Opportunity*, Collins, 2007, p13

50. Want to Know, 'Car mileage: 1908 Ford Model T – 25 MPG, 2004 EPA Average All Cars 21MPG' www.wanttoknow.info/050711carmileageaveragempg cited in George Monbiot, *Bring on the Apocalypse: Six Arguments for Global Justice*, Atlantic Books, 2008, p6

51. Lux Research, *The Cleantech Report*, 2007

52. George Bush, State of the Union Address, 28 January 2003. http://www.whitehouse.gov/news/releases/2003/01/20030128-19.html

53. US Department of Energy Hydrogen Program website, 'About page', http://www.hydrogen.energy.gov/about.html viewed 13/3/08

54. US Department of Energy Hydrogen Program website, 'Timeline', http://www.hydrogen.energy.gov/timeline.html viewed 13/3/08

55. European Commission Press Release, 'The Fuel Cells and Hydrogen Joint Technology Initiative', 10 October 2007 http://europa.eu/rapid/pressReleasesAction.do?reference=MEMO/07/404&format=HTML&aged=0&language=EN&guiLanguage=en

56. IPCC, *Special Report on Carbon Dioxide Capture and Storage*, Cambridge University Press, 2005, p131. http://arch.rivm.nl/env/int/ipcc/pages_media/SRCCS-final/IPCCSpecialReportonCarbondioxideCaptureandStorage.htm

57. CO2 emissions from gasoline 8.8 kg/gallon. US Government Environmental Protection Agency Office of Transportation and Air Quality, *Emission Facts: Average Carbon Dioxide Emissions Resulting from Gasoline and Diesel Fuel* p2, February 2005. http://www.epa.gov/otaq/climate/420f05001.pdf

58. Wikipedia, Liquid Natural Gas, http://en.wikipedia.org/wiki/LNG, viewed 13/4/08

59. See box about BMW's Hydrogen 7 car

60. Nick Eyre, Malcolm Fergusson and Richard Mills, *Fuelling Road Transport: Implications for Energy Policy*, Energy Saving Trust, Institute of European Environment Policy and National Society for Clean Air report to the Department for Transport, 2002.November 2002, p35 http://www.environmental-protection.org.uk/assets/library/documents/FuellingRoadTransport-Fullreport.pdf
They state each kilowatt hour (kWh) of renewable electricity saves 100g/C per kWh if avoiding gas, compared to 60g/C per kWh if the renewable electricity is used to create hydrogen for use as a vehicle fuel. Note that's grammes of carbon, so multiply by 3.667 to get grammes of CO2. Hydrogen therefore saves 220kg/CO2 per MWh.
They state displacing gas saves 'approximately 100kgC' (or 367kg/CO2) because gas emits 370kg/CO2 per MWh. Coal emits 890kg/CO2 per MWh (Department for Business, Enterprise and Regulatory Reform, Fuel Mix Disclosure Data Table 2006-07. http://www.berr.gov.uk/energy/policy-strategy/consumer-policy/fuel-mix/page21629.html)

61. Tyndall Centre for Climate Change Research, *Decarbonising the UK – Energy for a Climate Conscious Future*, 2005, p74 http://www.tyndall.ac.uk/media/news/tyndall_decarbonising_the_uk.pdf

62. United States Department of Energy, *National Hydrogen Energy Roadmap*, November 2002, p11 http://www1.eere.energy.gov/hydrogenandfuelcells/pdfs/national_h2_roadmap.pdf

63. Joseph Romm, *The Hype About Hydrogen*, Island Press, 2003, p9

64. Petrol contains 31.1 megajoules/litre [George Thomas, *Overview of Storage Development - DOE Hydrogen Program*, Sandia National Laboratories for US Government Department of Energy, May 2000, p4, www1.eere.energy.gov/hydrogenandfuelcells/pdfs/storage.pdf] Page 2 of the same report says 'At ambient conditions (300 K, 1 atm.): the energy content of 1 liter of H2 is only 10.7kJ, three orders of magnitude too low for practical applications.' So uncompressed hydrogen contains one-2,907th of the energy of petrol, to be exact.

65. US Department of Energy, Hydrogen Storage Basics, http://www1.eere.energy.gov/hydrogenandfuelcells/education/basics_storage.html, viewed 13/4/08

66. US National Academy of Engineering Board on Energy and Environmental Systems, *The Hydrogen Economy: Opportunities, Costs, Barriers, and R&D Needs*, 2004, p38

67. Raymond Drnevich of major American hydrogen supplier Praxair, *Hydrogen Delivery: Liquefaction & Compression*, May 2003, p8 http://www1.eere.energy.gov/hydrogenandfuelcells/pdfs/liquefaction_comp_pres_praxair.pdf

68. UK electricity supply CO2 emissions are currently 480 g/kWh [Department for Business Enterprise and Regulatory Reform, Fuel Mix Disclosure Data Table, table 3, 2006-07 http://www.berr.gov.uk/energy/policy-strategy/consumer-policy/fuel-mix/page21629.html]. 480g/CO2 per kWh x 12.5-15kWh = 6-7.2kg/CO2 per kg hydrogen liquefied.

69. 1 (US) gallon of petrol contains 121.3 megajoules, 1 kilogram of hydrogen contains 120.1-141.9 megajoules. [US National Academy of Engineering Board on Energy and Environmental Systems, *The Hydrogen Economy: Opportunities, Costs, Barriers, and R&D Needs*, 2004, Appendix H, p240. http://books.nap.edu/openbook.php?record_id=10922&page=240] (By this figure, 1 UK gallon of petrol contains 145.7 megajoules)

70. CO2 emissions from gasoline 8.8kg/US gallon. [US Government Environmental Protection Agency Office of Transportation and Air Quality, *Emission Facts: Average Carbon Dioxide Emissions Resulting from Gasoline and Diesel Fuel* p2, February 2005, http://www.epa.gov/otaq/climate/420f05001.pdf]

71. Petrol contains 31.1 megajoules/litre [George Thomas, op. cit., footnote 64]. Hydrogen gas at 3600psi delivers 2mj/litre [George Thomas, op. cit. p6, footnote 64]. At 10,000psi (the maximum currently possible for any storage canister, according to the US Government [Department of Energy website 'Basics of Hydrogen Storage' http://www1.eere.energy.gov/hydrogenandfuelcells/education/basics_storage.html viewed 23/3/08]) it would be 5.5mj/litre.

72. Bossel & Eliasson, *Energy and the Hydrogen Economy*, 2003, p19, fig.10 http://www.hyweb.de/News/Bossel-Eliasson_2003_Hydrogen-Economy.pdf

73. Alec Brooks, 'CARB's Fuel Cell Detour on the Road to Zero Emission Vehicles', *Electric Vehicle World*, 7 May 2004 http://www.evworld.com/article.cfm?storyid=691

74. 'The BMW Hydrogen', BMW press release, 14 May 2007. http://www.pressclub.bmw.de/pressclub/gb02.nsf

75. Bruce Gain, 'Road Testing BMW's Hydrogen 7', *Wired*, 13 Nov 2006, http://www.wired.com/cars/energy/news/2006/11/72100

76. 'Hell and Hydrogen', *Technology Review*, MIT, March 2007 http://www.technologyreview.com/Energy/18301/page2/

77. ibid., http://www.technologyreview.com/Energy/18301/page1/

78. Marianne Mintz et al, *Cost of Some Hydrogen Fuel Infrastructure Options*, Argonne National Laboratory Transportation Technology R&D Center, January 2002 http://www.transportation.anl.gov/pdfs/AF/224.pdf

79. Don Huberts, testimony to House Science Committee, 'The Path To A Hydrogen Economy', 5 March 2003 http://gop.science.house.gov/hearings/full03/mar05/huberts.htm

80. Chester Dawson, 'Fuel Cells: Japan's Carmakers Are Flooring It', *Business Week*, 23 December 2002 http://www.businessweek.com/magazine/content/02_51/b3813084.htm

81. Dr Sukhvinder Badwal, 'Fuel cells', *Science on the way to the hydrogen economy*, Australian Academy of Science, 5 May 2006 http://www.science.org.au/sats2006/badwal.htm viewed 15/6/08

82. Marianne Mintz et al, *Cost of Some Hydrogen Fuel Infrastructure Options*, Argonne National Laboratory Transportation Technology R&D Center, January 2002, http://www.transportation.anl.gov/pdfs/AF/224.pdf

83. 'Looking Ahead: Fuel Producers Weigh in on Hydrogen's Fit in Cleaner Energy Production', *Fuel Cell Industry Report*, January 2003

84. NASA, Office of Safety and Mission Assurance, *Safety Standard for Hydrogen and Hydrogen Systems*, 1997, paragraph 601b(4), http://www.hq.nasa.gov/office/codeq/doctree/canceled/871916.pdf

85. Arthur D. Little, Inc. for US Department of Energy. *Guidance for Transportation Technologies: Fuel Choice for Fuel Cell Vehicles*, Final Report Phase Two, February 2002, appendix p107, http://www1.eere.energy.gov/hydrogenandfuelcells/pdfs/fuel_choice_fcvs.pdf

86. ibid.

87. James Hansel of Air Products and Chemicals Inc, 'Safety Considerations for Handling Hydrogen', presentation to the Ford Motor Company, Allentown, Pennsylvania, 12 June 1998

88. 'Toyota Recalls Fuel-Cell Cars Due to Hydrogen Leak', Agence France-Presse, 20 May 2003

89. Bossel & Eliasson, *Energy and the Hydrogen Economy*, 2003, p19 http://www.hyweb.de/News/Bossel-Eliasson_2003_Hydrogen-Economy.pdf

90. Jim Campbell, 'Hydrogen Delivery Technologies and Systems: Pipeline Transmission of Hydrogen', presentation to US Department of Energy, Strategic Initiatives for Hydrogen Delivery Workshop May 7-8, 2003, p6 http://www1.eere.energy.gov/hydrogenandfuelcells/pdfs/hydrogen_pipelines_pres_air_liquid.pdf

91. Fabien Boudjemaa,' The Onboard Reformer: A Transitional Solution?', *CLEFS CEA* no.50-51, winter 2004-2005, p35 http://www.cea.fr/var/cea/storage/static/gb/library/Clefs50/pdf/034a036junker-gb.pdf

92. Weiss, M., Heywood, J., Schafer, A., Natarajan, V. 'Comparative Assessment of Fuel Cell Cars', *Report LFEE 2003-001 RP*, MIT, 2003. http://lfee.mit.edu/publications/PDF/LFEE_2003-001_RP.pdf

93. 'First Shell-Branded Hydrogen station opened', Shell Press Release, 24 April 2003 http://www.shell.com/home/content/media-en/news_and_library/press_releases/2003/first_shell_branded_hydro_station_24042003.html

94. Shell International, *Energy Needs, Choices and Possibilities: Scenarios to 2050*, 2001, page 48 http://www.cleanenergyfunds.org/CaseStudies/Shell_2050.pdf

95. ibid.

96. 'Real energy solutions for the real world', Shell Press Release, 1 May 2007, http://www.shell.com/home/content/media-en/news_and_library/press_releases/2007/brand_campaign_01032007.html

97. J. Levene, B. Kroposki, and G. Sverdrup, *Wind Energy and Production of Hydrogen and Electricity - Opportunities for Renewable Hydrogen*, US National Renewable Energy Laboratory, March 2006, p2. http://www.nrel.gov/docs/fy06osti/39534.pdf

98. 'The BMW Hydrogen 7', BMW press release, 14 May 2007 http://www.pressclub.bmw.de/pressclub/gb02.nsf

99. Department of Business, Enterprise and Regulatory Reform, 'Fuel Mix Disclosure Data Table 2006-07', Table 3, http://www.dti.gov.uk/energy/policy-strategy/consumer-policy/fuel-mix/page21629.html

100. IPCC, *Special Report on Carbon Dioxide Capture and Storage*, Cambridge University Press, 2005 p131

101. Raymond Drnevich of major American hydrogen supplier Praxair, *Hydrogen Delivery: Liquefaction & Compression*, May 2003, p8 http://www1.eere.energy.gov/hydrogenandfuelcells/pdfs/liquefaction_comp_pres_praxair.pdf

102. BMW website 'Prices and Specifications: 750i', http://www.bmw.co.uk/bmwuk/pricesandspecifications_technical_specs/0,,1156_149348629__bs-Nw%3D%3D%40bb-N0xJ%40bm-SEw4Mg%3D%3D%40sit-bmwuk,00.html]

103. Jeffrey S. Dukes, 'Burning Buried Sunshine: Human Consumption Of Ancient Solar Energy', *Climatic Change* 61, 2003, p31-44.

104. ibid.

105. Department for Transport, 'RTFO FAQ', http://www.dft.gov.uk/pgr/roads/environment/rtfo/faq viewed 12/3/08

106. Department for Transport Press Release, 'Kelly announces review of indirect impacts of biofuel production', 21 February 2008, http://www.gnn.gov.uk/environment/fullDetail.asp?ReleaseID=354445&NewsAreaID=2

107. European Union Directive, 'Promotion of the use of biofuels and other renewable fuels for transport, Directive 2003/30/EC of the European Parliament and of the Council of 8 May 2003 on the promotion of the use of biofuels and other renewable fuels for transport' , *Official Journal of the European Union*, L123/42, 17 May 2003. http://ec.europa.eu/energy/res/legislation/doc/biofuels/en_final.pdf

108. European Commission, 'Proposal for a Directive of the European Parliament and of the Council on the promotion of the use of energy from renewable sources', January 2008, http://www.euractiv.com/29/images/renewable%20directive%20jan%202008_tcm29-169463.pdf

109. Cited in EurActive News, 'Commission scientists blast EU biofuels policy', 18 January 2008, http://www.euractiv.com/en/transport/commission-scientists-blast-eu-biofuels-policy/article-169668 viewed 17/2/08

110. Roger Harrabin, 'EU rethinks biofuel guidelines', *BBC News*, 14 January 2008 http://news.bbc.co.uk/1/hi/world/europe/7186380.stm viewed 2/2/08

111. Biofuel Watch et. al., *Agrofuels: Towards a reality check in nine key areas*, June 2007. http://www.tni.org/reports/ctw/agrofuels.pdf

112. Jean Ziegler, *Report of the Special Rapporteur on the right to food to the sixty second session of the United Nations General Assembly*, 22 August 2007, http://daccessdds.un.org/doc/UNDOC/GEN/N07/487/05/PDF/N0748705.pdf?OpenElement

113. For a full critique see: Biofuel Watch et. al., *Agrofuels: Towards a reality check in nine key areas*, June 2007. http://www.tni.org/reports/ctw/agrofuels.pdf

114. Timothy Searchinger et. al.,'Use of U.S. Croplands for Biofuels Increases Greenhouse Gases Through Emissions from Land Use Change', *Science* (10.1126/science.1151861), 7 February 2008 and Joseph Fargione et. al., 'Land Clearing and the Biofuel Carbon Debt', *Science* (10.1126/science.1152747), 7 February 2008

115. David Adam, 'Kelly orders biofuels review', *The Guardian*, 21 February 2008

116. Wetland International, 'Biofuel Less Sustainable than realised' http://www.wetlands.org/news.aspx?ID=804eddfb-4492-4749-85a9-5db67c2f1bb8 viewed 2/2/08

117. Five million hectares of largely forest land in ten months. J. G Goldammer, J.G. & A.A. Hoffmann, 'The forest fire situation in Indonesia' *Int. Forest Fire News* 26, FAO Corporate Document Repository, 2001 www.fao.org/DOCREP/006/AD653E/ad653e45.htm

118. Page et al estimate that between 0.81 and 2.57Gt of carbon were released to the atmosphere in 1997 as a result of burning peat and vegetation in Indonesia, a volume equivalent to 13–40% of the mean annual global carbon emissions from fossil fuels. S Page, F Siegert, J O Rieley, H D V Boehm, A Java, S Limin, 'The amount of carbon released from peat and forest fires in Indonesia during 1997' *Nature* 420 (6911), 2002, p61

119. 'Oilseeds, World Markets and Trade', United States Department of Agriculture Foreign Agricultural Service *Circular Series FOP 08-07*, August 2007

120. Greenpeace SE Asia, 'Indonesia makes it to 2008 Guinness World Records as fastest forest destroyer on the planet', 3 May 2007 www.greenpeace.org/seasia/en/news/indonesia-makes-it-to-2008-gui.
Also; Nellemann, C, L Miles, BP Kaltenborn, M Virtue, and H Ahlenius (Eds), The last stand of the orangutan – State of emergency: Illegal logging, fire and palm oil in Indonesia's national parks United Nations Environment Programme, 2007, www.unep.org/grasp/docs/2007Jan-LastStand-of-Orangutan-report.pdf

121. Marianne Klute, *Green Gold Biodiesel: Players in Indonesia*, Watch Indonesia, 2007 http://www.biofuelwatch.org.uk/Green_Gold_Biodiesel_%20Players_in_Indonesia.doc

122. W. H. Schlesinger, *Biogeochemistry: An Analysis of Global Change* (2nd Edition), San Diego: Academic Press, 1997, p588.

123. Joseph Fargione et. al., 'Land Clearing and the Biofuel Carbon Debt', *Science* (10.1126/science.1152747), 7 February 2008

124. Julia King, *The King Review of low-carbon cars Part I: the potential for CO2 reduction,* UK Government HM Treasury October 2007 www.hm-treasury.gov.uk/king

125. Timothy Searchinger et al.,'Use of U.S. Croplands for Biofuels Increases Greenhouse Gases Through Emissions from Land Use Change', *Science* (10.1126/science.1151861), 7 February 2008

126. Greenpeace International, *Cooking The Climate*, November 2007, p52. http://www.greenpeace.org/raw/content/international/press/reports/cooking-the-climate-full.pdf

127. UN Food and Agriculture Organisation, *Biofuels and Commodity Markets – Palm Oil Focus*, 2006 http://www.fao.org/es/ESC/common/ecg/110542_en_full_paper_English.pdf

128. UN Food and Agriculture Organisation. *World agriculture: towards 2030/2050. Prospects for food, nutrition, agriculture and major commodity groups*, Global Perspective Studies Unit, FAO, June 2006

129. Julian Borger, 'Feed the world? We are fighting a losing battle, UN admits', *The Guardian*, 26 February 2008 http://www.guardian.co.uk/environment/2008/feb/26/food.unitednations

130. Lian Pin Koh, 'Potential Habitat and Biodiversity Losses from Intensified Biodiesel Feedstock Production', *Conservation Biology*, Vol. 21 Issue 5 Page 1373, October 2007

131. Global Nature Fund and Ecotropica, 'Pantanal in South America declared "Threatened Lake of the Year 2007"', 1st February 2007 http://www.ramsar.org/wwd/7/wwd2007_rpts_germany_gnf.htm viewed 17/2/08

132. Bruno Waterfield and Charles Clover, 'Set aside suspended by European Union', *Daily Telegraph*, 26 September 2007

133. For example the 'Alert Against the Green Desert Network Manifesto' from a coalition of Brazillian civil society groups opposing expansion of monoculture tree plantations. http://www.wrm.org.uy/countries/Brazil/manifesto.html

134. For a full analysis see: GRAIN, 'Against the Grain: Sustainable Monoculture? No Thanks!', June 2006, http://www.grain.org/articles_files/atg-6-en.pdf and Biofuel Watch et. al.
Also; *Agrofuels: Towards a reality check in nine key areas*, BiofuelWatch et al, June 2007 http://www.tni.org/reports/ctw/agrofuels.pdf

135. BP began this with the purchase of Solarex in 1999, ('BP Amoco Plans to Buy Remaining 50% Stake in Solarex', *New York Times*, 7 April 1999), Shell also has investments in solar power, though most of these holdings were sold off in 2007 (Terry Macalister, 'Big Oil lets the sun set on renewables', *The Guardian*, 11 December 2007).

136. US Department of Energy, 'Fuel Ethanol Production', Genomics to Life Programme, http://genomicsgtl.energy.gov/biofuels/ethanolproduction.shtml

137. 'Cellulose Ethanol is Ready to Go: Iogen producing world's first cellulose ethanol fuel', Iogen press release, 21 April 2004. http://www.iogen.ca/news_events/press_releases/2004_04_21.html

138. LS9 website, 'Renewable Petroleum Technology', http://www.ls9.com/technology/ viewed 13/3/08

139. Mark Clayton, 'US to cut funds for two renewable energy sources', *Christian Science Monitor*, 15 September 2006 http://www.csmonitor.com/2006/0915/p02s01-uspo.html

140. Brown, L.R., 'World's rangelands deteriorating under mounting pressure', Earth Policy Institute, 5 February 2002, http://www.earth-policy.org/Updates/Update6.htm.

141. Department of Energy and Department of Agriculture, *Biomass as a feedstock for the bioenergy and bioproducts industry: The technical feasibility of a billion tonne*

annual supply, April 2005. http://feedstockreview.ornl.gov/pdf/billion_ton_vision.pdf

142. Alice Friedemann, 'Peak Soil: Why Cellulosic ethanol and other Biofuels are Not Sustainable and are a Threat to America's National Security', April 10, 2007, http://www.culturechange.org/cms/index.php?option=com_content&task=view&id=107&Itemid=1 viewed 10/3/08

143. For more see: ETC Group, *Peak Soil + Peak Oil = Peak Spoils*, November/December 2007, http://www.etcgroup.org/en/materials/publications.html?pub_id=668

144. ETC Group, ibid., and Biofuel Watch et. al., *Agrofuels: Towards a reality check in nine key areas*, June 2007 http://www.tni.org/reports/ctw/agrofuels.pdf?

145. ibid.

146. ibid.

147. Purdue University news, 'Fast-growing GM trees could take root as future energy source', 23 August 2006 http://www.purdue.edu/UNS/html4ever/2006/060823.Chapple.poplar.html

148. See for more information Joe Cummins and Mae-Wan Ho, 'Moratorium on all GM Trees and Ban on GM Forest Trees', Institute of Science and Society, 12 February 2007, http://www.i-sis.org.uk/Moratorium_on_all_GM_Trees.php

149. Synthetic Genomics Press Release, 'Venter Institute Scientists Create First Synthetic Bacterial Genome', 24 January 2008, http://www.syntheticgenomics.com/press/2008-01-24.htm

150. ETC Group Press Release, 'Goodbye, Dolly...Hello, Synthia! J. Craig Venter Institute Seeks Monopoly Patents on the World's First-Ever Human-Made Life Form', 7 June 2007, http://www.etcgroup.org/en/materials/publications.html?pub_id=631

151. Synthetic Genomics Press Release, 'Venter Institute Scientists Create First Synthetic Bacterial Genome', 24 January 2008, http://www.syntheticgenomics.com/press/2008-01-24.htm

152. Cited in ETC Group Press Release, 'Goodbye, Dolly...Hello, Synthia! J. Craig Venter Institute Seeks Monopoly Patents on the World's First-Ever Human-Made Life Form'

153. US Department of Energy, News Release, 'Researchers Funded by the DOE 'Genomes to Life' Program Achieve Important Advance in Developing Biological Strategies to Produce Hydrogen, Sequester Carbon Dioxide and Clean up the Environment', 13 November 2003

154. Synthetic Genomics Press Release, 'Dr. Aristides Patrinos Named President Of Synthetic Genomics, Inc.', 2 February 2006, http://www.syntheticgenomics.com/press/2006-02-02.htm

155. Synthetic Genomics Press Release, 'Venter Institute Scientists Create First Synthetic Bacterial Genome', 24 January 2008, http://www.syntheticgenomics.com/press/2008-01-24.htm

156. Synthetic Genomics Press Release, 'Synthetic Genomics Inc. and BP to Explore Bioconversion of Hydrocarbons into Cleaner Fuels', June 13, 2007, http://www.syntheticgenomics.com/press/2007-06-13.htm

157. For a full analysis see ETC Group, *Extreme Genetic Engineering: An Introduction to Synthetic Biology*, January 2007, http://www.etcgroup.org/en/issues/synthetic_biology.html

158. ibid.

159. *Well-To-Wheels Analysis Of Future Automotive Fuels And Powertrains In The European Context: Well-to-Wheels Report version 1b*, Concawe/Eucar/JRC-IES, January 2004, p57 fig9.2 http://www.enpc.fr/fr/formations/dea_masters/tradd/documents/Welltowheeleucar2004.pdf

160. Cited in Fred Pearce, 'Biofuels: the future?', *Daily Telegraph*, 2 February 2008

161. European Commission, *Sustainable power generation from fossil fuels: aiming for near-zero emissions from coal after 2020*, 10 January 2007. http://ec.europa.eu/energy/energy_policy/doc/16_communication_fossil_fuels_en.pdf

162. Bert Metz, Ogunlade Davidson, Heleen de Coninck, Manuela Loos, Leo Meyer (Eds), *Carbon Dioxide Capture and Storage: Summary for Policymakers and Technical Summary, A report of Working Group III of the IPCC*, Cambridge University Press, 2005, p2 & p45

163. ibid.

164. ibid., p3

165. ibid., p4

166. ibid., p5-7

167. ibid.

168. ibid., p30

169. S. Solomon, *Carbon Dioxide Storage: Geological Security and Environmental Issues - Case Study on the Sleipner Gas Field in Norway*, Bellona, 2006. http://www.bellona.no/artikler/notat_solomon

170. Bert Metz, Ogunlade Davidson, Heleen de Coninck, Manuela Loos, Leo Meyer (eds), op. cit. (footnote 162), p13

171. ibid., p4

172. ibid.

173. ibid., p5

174. ibid.

175. ibid.

176. ibid.

177. ibid.

178. ibid., p6

179. Fred Pearce, 'Cleaning Up Coal', *New Scientist*, 29 March 2008

180. Bert Metz, Ogunlade Davidson, Heleen de Coninck, Manuela Loos, Leo Meyer (eds), op. cit. (footnote 162), p13

181. ibid., p12

182. ibid.

183. ibid., p7

184. Department for Business Enterprise and Regulatory Reform, 'CCS Demonstration Competition: Documentation/Q & A', http://www.berr.gov.uk/energy/sources/sustainable/carbon-abatement-tech/ccs-demo/docs-qa/page42503.html viewed 19/12/08

185. European Commission, *Sustainable power generation from fossil fuels: aiming for near-zero emissions from coal after 2020*, 10 January 2007 p4

186. ibid., p11

187. Department for Business Enterprise and Regulatory Reform, 'CCS Demonstration Competition', http://www.berr.gov.uk/energy/sources/sustainable/carbon-abatement-tech/ccs-demo/page40961.html

188. US Department of Energy, 'DOE Announces Restructured FutureGen Approach to Demonstrate Carbon Capture and Storage Technology at Multiple Clean Coal Plants', 30 January 2008 http://www.fossil.energy.gov/news/techlines/2008/08003-DOE_Announces_Restructured_FutureG.htmlviewed 19/2/08

189. OFGEM and Department for Business, Enterprise and Regulatory Reform, *Energy Markets Outlook: Chapter 6 – Coal*, October 2007, p63. http://www.berr.gov.uk/files/file42000.pdf

190. 'Drax buys more EU carbon permits in 2006', Reuters, 8 March 2007, http://uk.reuters.com/article/businessNews/idUKL0849005620070308

191. US Department of Energy, 'DOE Announces Restructured FutureGen Approach to Demonstrate Carbon Capture and Storage Technology at Multiple Clean Coal Plants'

192. FutureGen for Illinios, 'Project Timeline'. http://futuregenforillinois.org/projecttimeline.htm viewed 19/2/08

193. Bert Metz, Ogunlade Davidson, Heleen de Coninck, Manuela Loos, Leo Meyer (eds), op. cit. (footnote 162), p8

194. Fred Pearce, 'Cleaning Up Coal', *New Scientist*, 29 March 2008

195. Peter Viebahn, 'Comparison of Carbon Capture and Storage with Renewable Energy Technologies Regarding Structural, Economic, and Ecological Aspects',ACCSEPT workshop, Bonn, 10-11 May 2007, http://www.accsept.org/outputs/peter_viebahn_bonn.pdf

196. World Coal Institute, 'Coal Facts 2007' http://www.worldcoal.org/pages/content/index.asp?PageID=188 viewed 13/3/07

197. The last to be opened was Stage 2 of Drax, in 1986. Drax website, 'History', http://www.draxpower.com/aboutus/history/ viewed 13/02/08

198. 8.5GW of coal-fired capacity will close to meet the requirements of the EU Large Combustion Plant Directive (LCPD) by end 2015; as will about 2.5GW of oil power stations. Department of Trade and Industry, *Meeting the Energy Challenge*, p128-129

199. E.ON website, 'Kingsnorth Cleaner Coal Plant', http://www.eon-uk.com/generation/supercritical.aspx viewed 11/4/08

200. John Sauven 'Agreeing to Kingsnorth will signal surrender on targets', *The Independent*, 17 December 2007

201. Including Tilbury, Blythe, High Marnham, Ferrybridge, Longannet and Cockenzie. Department of Trade and Industry, 'First Meeting of the UK Coal Forum', 14th November 2006 http://www.berr.gov.uk/files/file37293.pdf

202. The Energy Review says: 'We are considering whether we should require future fossil fuel power stations to be built "capture ready" and if so what the options would be for doing this. We will need to recognise that for some projects the scope for CCS (eg because of geographical location or other technical limitations) may be limited. We will launch a consultation later this year on the aspects of capture readiness which should be included in future applications for consent and how practically we might deal with this issue in the consenting process'. Department for Business Enterprise and Regulatory Reform, 'Chapter 5: Electricity Generation', *Energy Challenge: Energy Review: A Report*, 11 July 2006, p177. http://www.berr.gov.uk/files/file39569.pdf

203. Greenpeace Press Release, 'Whitehall emails reveal government climate policy being dictated by German utility giant', 31 January 2008, http://www.greenpeace.org.uk/media/press-releases/government-climate-policy-dictated-by-german-utility-giant-20080131 viewed 17/3/08

204. E.ON Press Release, 'E.ON enters UK Government's carbon capture and storage competition', 31 March 2008, http://pressreleases.eon-uk.com/blogs//eonukpressreleases/archive/2008/03/31/E.ON-enters-UK-Government_2700_s-Carbon-Capture-and-Storage-competition.aspx

205. The government predict that adaptation of existing power stations would happen between 2016 and 2025. (OFGEM and Department for Business, *Enterprise and Regulatory Reform, Energy Markets Outlook: Chapter 6 – Coal*, October 2007, p64. http://www.berr.gov.uk/files/file42000.pdf). A decade is a reasonable assumption. These figures are however contradicted by the fact that the demonstration project will not be completed until 2020, and the EU estimate for when CCS will be economically competitive is 2020. So it is more likely that plants will be retrofitted for CCS between 2020 and 2030. The government acknowledge that some locations where coal power plants are being proposed may not be suitable for CCS.

206. Department for Business Enterprise and Regulatory Reform, 'Chapter 5: Electricity Generation', *Energy Challenge: Energy Review: A Report*, 11 July 2006, p177. http://www.berr.gov.uk/files/file39569.pdf

207. OFGEM and Department for Business, Enterprise and Regulatory Reform, Energy Markets Outlook: Chapter 6 – Coal, October 2007, p63. http://www.berr.gov.uk/files/file42000.pdf p11

208. For more information of Ffos-y-Fran and campaigning against it see The Coal Hole website http://thecoalhole.org

209. Miller Argent, Ffos-y-Fran Land Reclamation Scheme website, 'Coal Sales', http://www.ffos-y-fran.co.uk/index.cfm?id=8

210. George Monbiot, 'The New Coal Age', *The Guardian*, 9 October 2007

211. Miller Argent, Ffos-y-Fran Land Reclamation Scheme website, http://www.ffos-y-fran.co.uk/

212. Miller Argent, Ffos-y-Fran Land Reclamation Scheme website, 'Coal Sales', http://www.ffos-y-fran.co.uk/index.cfm?id=8

213. Figures by George Monbiot in 'The New Coal Age', *The Guardian*, 9 October 2007. Based on one tonne of coal containing 746 kg of carbon, burning it produces 2.7 tonnes of carbon dioxide since CO2 is 3.667 times the molecular weight of C. Figure for carbon content of coal from: Bioenergy Feedstock Information Network, 'Bioenergy Conversion Factor', http://bioenergy.ornl.gov/papers/misc/energy_conv.html viewed 17/3/08

214. British Geological Survey, 'Coal: opencast coal mining statistics 2006,' Table 4, 2007. http://www.bgs.ac.uk/mineralsuk/minequar/coal/occ/home.html#TABLE4

215. Department for Business Enterprise and Regulatory Reform page, 'UK Coal Forum' http://www.berr.gov.uk/energy/sources/coal/forum/page37276.html viewed 17/3/08

216. BP Press Release, 'BP And Partners Plan Clean Energy Plant in Scotland, Increasing Oil Recovery And Reducing Emissions', 30 June 2005 http://www.bp.com/genericarticle.do?categoryId=2012968&contentId=7006999

217. ibid. BP expected to store 1.3million tonnes of carbon dioxide per year for 20 years.

218. The combined liquid fuels from an average barrel of crude oil will produce a minimum of 317kg of CO2 when consumed. Jim Bliss, 'Carbon dioxide emissions per barrel of crude', The Quiet Road, 20 March 2008, http://numero57.net/?p=255 viewed 27/03/08

219. See for example: David Fleming, *The Lean Guide to Nuclear Energy*, The Lean Economy Connection, 2007 http://www.theleaneconomyconnection.net/nuclear/index.html.
Also; Chris Grimshaw and Olaf Bayer, *Broken Promises : Why The Nuclear Industry Won't Deliver,* Corporate Watch, 2007 http://www.corporatewatch.org.uk/download.php?id=68
Also; Friends of the Earth, *Nuclear Power, Climate Change and the Energy Review,* June 2006 http://www.foe.co.uk/resource/briefings/nuclear_power.pdf.
Also; Greenpeace, *Nuclear Power: Undermining Action on Climate Change*, December 2007, http://www.greenpeace.org/raw/content/international/press/reports/nuclear-power-undermining-ac.pdf

220. For examples visit 'Nuclear Spin' and 'Spinwatch' websites: www.nuclearspin.org and www.spinwatch.org

221. Iain MacLeay, Kevin Harris, Chris Michaels, *Digest of United Kingdom Energy Statistics*, National Statistics/Department for Business, Enterprise & Regulatory Reform, 2007, p118. http://stats.berr.gov.uk/energystats/dukes07.pdf

222. Department for Business Enterprise and Regulatory Reform, *Energy Challenge: Energy Review: A Report*, July 2006, http://www.berr.gov.uk/files/file31890.pdf

223. 'Nuclear review "was misleading"', *BBC News,* 15 February 2007 http://news.bbc.co.uk/1/hi/uk_politics/6364281.stm

224. Department for Business Enterprise and Regulatory Reform, *Meeting the energy challenge: a white paper on nuclear power*, January 2008, http://www.berr.gov.uk/energy/nuclear-whitepaper/page42765.html viewed 13/3/08

225. John Vidal, 'Consumers may foot nuclear bill', *The Guardian*, 7 January 2008

226. Chris Grimshaw and Olaf Bayer, *Broken Promises : Why The Nuclear Industry Won't Deliver*, Corporate Watch, 2007 http://www.corporatewatch.org.uk/download.php?id=68

227. See for example David Fleming, *The Lean Guide to Nuclear Energy*, The Lean Economy Connection, 2007, p9 http://www.theleaneconomyconnection.net/nuclear/index.html

228. The Consultation accepts estimates derived from the Sustainable Development Commission (2006), 'Chapter 2', *The Role of Nuclear Power in a Low Carbon Economy*, p22

229. ISA, *Life-Cycle Energy Balance and Greenhouse Gas Emissions of Nuclear Energy in Australia*, University of Sydney, 2005, at http://www.pmc.gov.au/publications/umpner/docs/commissioned/ISA_report.pdf

230. ibid. p7 & p171

231. J. W. Storm van Leeuwen, 'Greenhouse Gases from Nuclear', *Appendix B, in Evidence to the IPCC Working Group III, Fourth Assessment Report*, 2006

232. Sustainable Development Commission, *The role of nuclear power in a low carbon economy*, March 2006 p6

233. For full analysis see, Chris Grimshaw and Olaf Bayer, *Broken Promises : Why The Nuclear Industry Won't Deliver*, Corporate Watch, 2007 http://www.corporatewatch.org.uk/download.php?id=68

234. Greenpeace, *Nuclear Power: Undermining Action on Climate Change*, December 2007 http://www.greenpeace.org/raw/content/international/press/reports/nuclear-power-undermining-ac.pdf

235. For a full analysis see: Chris Grimshaw and Olaf Bayer, *Broken Promises : Why The Nuclear Industry Won't Deliver*, Corporate Watch, 2007

236. David Fleming, *The Lean Guide to Nuclear Energy*, The Lean Economy Connection, 2007 http://www.theleaneconomyconnection.net/nuclear/index.html p9

237. Terry Macalister, 'Minister admits nuclear fuel plant produces almost nothing', *The Guardian*, 3rd March 2008 http://www.guardian.co.uk/world/2008/mar/03/nuclear.energy

238. Committee on Radioactive Waste Management, 'Final Report', Chapter 16, paragraph 55, p132 http://www.corwm.org/pdf/FullReport.pdf

239. David Fleming, *The Lean Guide to Nuclear Energy*, The Lean Economy Connection, 2007 http://www.theleaneconomyconnection.net/nuclear/index.html p9

240. ibid.

241. Terry MacAlister, 'Ministers 'wrote blank cheque' to bail out nuclear power group', *The Guardian*, 17 March 2006

242. 'More a question of politics than economics', *Q Series, The Future of Nuclear,* UBS Investment Research, 2005

243. 'Nuclear Clean Up Cost: 2020 Fission: A financial Analysis of Nuclear Decommissioning Costs', HSBC Global, September 2005

244. National Decommissioning Authority, *NDA Strategy*, 30 March 2006, p77

245. John Vidal, 'Consumers may foot nuclear bill', *The Guardian*, 7 January 2008

246. John Vidal, 'Nuclear Plants Bloom', *The Guardian*, 12 August 2004

247. Cited in Commons debate, 22 January 2008, Hansard column 1395, http://www.publications.parliament.uk/pa/cm200708/cmhansrd/cm080122/debtext/80122-0010.htm (citation: *Official Report*, 24 February 2003; Vol. 400, c. 32)

248. Flood Hazard Research Centre and Greenpeace, *The impacts of climate change on nuclear power stations sites a review of four proposed new-build sites on the UK coastline*, March 2007

249. Greenpeace, 'Submissions to Environment Agency, Comments on Environment Agency's Assessment Documents on Drigg', September 2005, http://www.greenpeace.org.uk/MultimediaFiles/Live/FullReport/7306.pdf viewed 12/2/08

250. 'France gets nuclear fusion plant', *BBC News,* 28 June 2005, http://news.bbc.co.uk/1/hi/sci/tech/4629239.stm viewed 19/1/08

251. Greenpeace, ESTIA and Solar Paces, *Concentrated Solar Thermal Power - Now*, September 2005

252. FPL Energy website, SEGS III, IV, V, VI, VII, VIII & IX, http://www.fplenergy.com/portfolio/contents/segs_viii.shtml viewed 14/2/08

253. Matthias Fawer, 'Solar Energy 2007', *London Accord*, Sarasin and London Accord, November 2007, http://www.london-accord.co.uk/final_report/reports/pdf/c1.pdf

254. Derived from Iain MacLeay, Kevin Harris, Chris Michaels, *Digest of United Kingdom Energy Statistics*, 'Chapter 7: Renewable sources of energy', National Statistics/ Department for Business, Enterprise and Regulatory Reform, 2007. http://stats.berr.gov.uk/energystats/dukes07.pdf

255. Matthias Fawer, 'Solar Energy 2007', *London Accord*, Sarasin and London Accord, November 2007, http://www.london-accord.co.uk/final_report/reports/pdf/c1.pdf

256. Subhashish Banerjee, L.J. Duckers1, R. Blanchard, and B.K. Choudhury, 'Life Cycle Analysis of Selected Solar and Wave Energy Systems', *Advances in Energy Research*, 2006

257. See: Energy Technology Support Unit, *New and Renewable Energy: Prospects in the UK for the 21st Century – Supporting Analysis*, Harwell, 1999 p133.
Also; Jim Watson, 'Co-Provision in UK Sustainable Energy Systems: The Case of Microgeneration', *Energy Policy*, Vol 32 (March 2004) p1981-1990.
Also; Royal Commission on Environmental Pollution, *Energy: The Changing Climate*, June 2003, para 7.40www.rcep.org.uk/newenergy.htm

258. Jeremy Leggett, 'Keep on the sunny side', *Comment is Free*, 18 December 2007 http://www.guardian.co.uk/commentisfree/2007/dec/18/keeponthesunnyside

259. A. Klein, et al, *Evaluation of different feed-in tariff design options best practice paper for the International Feed In cooperation*, Frounhofer Institute Systems and Innovation Research, June 2007 http://www.feed-incooperation.org/images/files/best_practice_paper_final.pdf

260. N. Stern, *Stern Review – The Economics of Climate Change,* Part IV 'Policy Responses for Mitigation', chapter 16, HM Treasury, 2006. http://www.hm-treasury.gov.uk/media/9A3/57/Ch_16_accelerating_technological_innovation.pdf

261. CAES Development Company, 'The CEAS [Compressed Air Energy Storage] Cycle', http://www.caes.net/cycle.html viewed 19/2/08

262. G. Phylipsen, and E.Alsema, *Environmental life-cycle assessment of multicrystalline silicon solar cell modules*, Department of Science, Technology and Society, Utrecht University, September 1995 http://www.chem.uu.nl/nws/www/publica/95057.pdf

263. ibid.

264. Nanosolar website, 'Nanoparticle Ink', http://www.nanosolar.com/nanoink.htm viewed 13/3/08

265. See for more information: Olaf Bayer, *Nanomaterials: Undersized, unregulated and already here*, Corporate Watch, 2007. http://www.corporatewatch.org/download.php?id=63

266. G. Phylipsen, and E.Alsema, op. cit. (footnote 262)

267. Vestas brochure, 'V90-3.0 MW: An efficient way to more power', September 2007

268. GWEC, Greenpeace International, DLR and Ecofys, Global Wind Energy Outlook 2006, September 2006. http://www.gwec.net/fileadmin/documents/Publications/GWEC_A4_0609_English.pdf

269. ibid., p5

270. ibid., p10

271. Department for Business, Enterprise and Regulatory Reform website, 'What is the Renewables Obligation?', http://www.berr.gov.uk/energy/sources/renewables/policy/renewables-obligation/what-is-renewables-obligation/page15633.html viewed 13/3/08

272. HM Revenue and Customs website, 'The Climate Change Levy: Introduction', http://customs.hmrc.gov.uk/channelsPortalWebApp/channelsPortalWebApp.portal?_nfpb=true&_pageLabel=pageExcise_InfoGuides&propertyType=document&id=HMCE_CL_001174#P1_34 viewed 13/3/08

273. The corporate front groups that has lead this attack is Supporters of Nuclear Energy (SONE). More information is available at the Nuclear Spin website, 'Supporters of Nuclear Energy', www.nuclearspin.org/index.php/Supporters_of_Nuclear_Energy#Recent_Pro-Nuclear_and_Anti-Wind_activities_of_SONE, viewed 12/3/08

274. See section on 'optimised integration of renewable energy' in Greenpeace International, European Renewable Energy Council (EREC), *Energy [R]evolution: A sustainable world energy outlook*, January 2007, p21-22

275. Greenpeace International, European Renewable Energy Council (EREC), *Energy [R]evolution: A sustainable world energy outlook*, January 2007, p22

276. ibid., p21-22

277. There has been opposition to a number of windfarms in the UK for sound environmental reasons such as Lewis (peat), Cefn Croes (peat and tree felling).

278. Ed Douglas, 'The hidden cost of wind', *New Scientist*, 8 July 2006, http://technology.newscientist.com/article/mg19125591.600

279. See for example Royal Society for the Protection of Birds, Wind Farms, http://www.rspb.org.uk/ourwork/policy/windfarms/index.asp viewed 13/4/08

280. British Wind Energy Association, 'Offshore Wind: Round One Wind Farms', www.bwea.com/offshore/round1.html viewed 19/02/08

281. Department for Business, Enterprise and Regulatory Reform website, 'Wave Energy: How it Works', www.berr.gov.uk/energy/sources/renewables/explained/wave-tidal/wave/page17058.html viewed 19/2/08

282. Wikipedia, 'Small Scale Hydro', http://en.wikipedia.org/wiki/Small_hydro viewed 27/3/08

283. Scottish and Southern Energy, 'Glendoe Hydro Scheme, Title Page', www.glendoe.co.uk viewed 19/2/08

284. Derived from Iain MacLeay, Kevin Harris, Chris Michaels, Digest of United Kingdom Energy Statistics, 'Chapter 7: Renewable sources of energy', National Statistics/ Department for Business, Enterprise and Regulatory Reform, 2007. http://stats.berr.gov.uk/energystats/dukes07.pdf

285. For a full analysis see: *Dams and Development: A New Framework for Decision Making*, World Commission on Dams, 2000. Particularly chapter 3 (Ecosystems and Large Dams: Environmental Performance) and chapter 4 (People and Large Dams: Social Performance)

286. E Duchemin, M Lucotte, VL St.Louis, R Canuel. 'Hydroelectric reservoirs as an anthropogenic source of greenhouse gases' , *World Resource Review* 14, 2002, p334-353

287. Forster, P., et al V. Ramaswamy, P. Artaxo, T. Berntsen, R. Betts, D.W. Fahey, J. Haywood, J. Lean, D.C. Lowe, G. Myhre, J. Nganga, R. Prinn, G. Raga, M. Schulz and R. Van Dorland, 2007: Changes in Atmospheric Constituents and in Radiative Forcing. *Climate Change 2007: The Physical Science Basis. Contribution of Working Group I to the Fourth Assessment Report of the Intergovernmental Panel on Climate Change*, p.212, Table 2.14. [Solomon, S., D. Qin, M. Manning, Z. Chen, M. Marquis, K.B. Averyt, M.Tignor and H.L. Miller (eds.)]. Cambridge University Press, Cambridge, UK and New York, NY, USA. http://ipcc-wg1.ucar.edu/wg1/Report/AR4WG1_Print_Ch02.pdf

288. E Duchemin, M Lucotte, VL St.Louis, R Canuel. 'Hydroelectric reservoirs as an anthropogenic source of greenhouse gases.' *World Resource Review* 14, 2002, p334-353

289. Philip M. Fearnside, 'Do Hydroelectric Dams Mitigate Global Warming? The Case of Brazil's Curuá-Una Dam', *Mitigation and Adaptation Strategies for Global Change* Volume 10 Number 4, October 2005

290. ibid.

291. *Dams and Development: A New Framework for Decision Making*, World Commission on Dams, 2000 p75

292. 'Water power 'fuels climate change'', *BBC News,* 31 May 2000, http://news.bbc.co.uk/1/hi/sci/tech/771465.stm viewed 19/2/08

293. These are: Rance River, France; Bay of Fundy, Canada; and Kislaya Guba, Russia

294. Wikipedia; 'Tidal Power', http://en.wikipedia.org/wiki/Tidal_power, viewed 27/03/08

295. ibid.

296. Department for Business, Enterprise and Regulatory Reform, 'Tidal Power; Severn Estuary, http://www.berr.gov.uk/energy/sources/renewables/explained/severntidalpower/page41473.html viewed 27/03/08

297. Sustainable Development Committee, 'Turning the Tide: Tidal Power in the UK', October 2007, p11

298. ibid. p7

299. ibid. p80–93 & 97

300. ibid. p100

301. ibid. p77-82

302. Ed Owen, 'Environment Agency boss slams Severn Barrage', *New Civil Engineer*, 7 November 2007. http://www.nce.co.uk/news/environment_agency_boss_slams_severn_barrage.html

303. Friends of the Earth, *A Severn Barrage or Tidal Lagoon*, January 2004, http://www.foe.co.uk/resource/briefings/severn_barrage_lagoons.pdf

304. Department for Business, Enterprise and Regulatory Reform, 'Wave Energy: Current Use', www.berr.gov.uk/energy/sources/renewables/explained/wave-tidal/wave/page17048.html viewed 19/2/08

305. ibid.

306. Greenpeace International, European Renewable Energy Council (EREC), *Energy [R]evolution: A sustainable world energy outlook*, January 2007, p7

307. Iain MacLeay, Kevin Harris, Chris Michaels, *Digest of United Kingdom Energy Statistics*, 'Chapter 7: Renewable sources of energy', National Statistics/ Department for Business, Enterprise and Regulatory Reform, 2007. http://stats.berr.gov.uk/energystats/dukes07.pdf

308. Department for Business, Enterprise and Regulatory Reform website, 'What is the Renewables Obligation?', http://www.berr.gov.uk/energy/sources/renewables/policy/renewables-obligation/what-is-renewables-obligation/page15633.html viewed 13/3/08

309. Department for Business, Enterprise and Regulatory Reform website, 'Environmental Transformation Fund (ETF): Funding for Low Carbon Technologies', http://www.berr.gov.uk/energy/environment/etf/page41652.html, viewed 25/3/08

310. Drax Press Release, 'Short rotation coppice willow processing trial proceeds to second phase', 5 September 2005. http://www.draxgroup.plc.uk/media/press_releases/?id=1438

311. Fred Pearce, 'Biofuels: the future?', *Daily Telegraph*, 2 February 2008 www.telegraph.co.uk/earth/main.jhtml?xml=/earth/2008/02/02/sm_biofuels02.xml&page=1

312. Drax burns around 10 million tonnes of coal a year [*Annual Report and Accounts 2006*, Drax Group plc, March 2007, p21 http://www.draxgroup.plc.uk/annual2006/files/page/5183/complete.pdf]. Drax estimate that it takes 1.5 times the amount of biomass to replace a given weight of coal ['Alstom to build £50m biomass plant for Drax', *The Guardian*, 20 May 2008], so 1.5 million tones of biomass is required. The willow grown for Drax's trial yielded just under 10 tonnes per hectare ['Drax Goes Green with Willow', *The Guardian*, 19 March 2004], meaning 150,000 hectares would be needed. Grown on three year rotation (the fastest possible) means 450,000 hectares would be required to supply Drax with 10% biomass.

313. Scottish Power Press Release, ScottishPower Announces UK's Largest Energy Crop Plan, 19 July 2007 http://www.manweb.com/PressReleases_1572.htm

314. ibid.

315. Prenergy Power, 'Homepage', http://www.prenergypower.com viewed 17/3/08

316. ibid.

317. David Williamson, 'Wood plant 'to power half of Wales'', *Western Mail*, 12 October 2006

318. E.ON, 'Biomass', http://www.eon-uk.com/generation/biomass.aspx viewed 17/3/07

319. Nigel Bruce, Rogelio Perez-Padilla, & Rachel Albalak, 'Indoor air pollution in developing countries: a major environmental and public health challenge', *Bulletin of the World Health Organization*, 2000, 78 (9), p1078

320. All refs; Friends of the Earth, *Dirty truths – Incineration and climate change,* May 2006 http://www.foe.co.uk/resource/briefings/dirty_truths.pdf

321. Greenpeace International, European Renewable Energy Council (EREC), *Energy [R]evolution: A sustainable world energy outlook*, January 2007 p21

322. ibid., p21-22

323. ibid., p20

324. Gerhard Knies, Uwe Möller, Michael Straub, *The DESERTEC Concept for Energy, Water and Climate Security: White Paper*, The Club of Rome and Trans-Mediterranean Renewable Energy Co-operation (TREC), November 2007. http://www.desertec.org/downloads/articles/trec_white_paper.pdf

325. DC or direct current, as opposed to alternating current which the UK grid currently runs on. Changes in the way electricity is generated, transmitted, and consumed may mean a shift to using DC current as standard.

326. DesertTec, 'About TREC',' http://www.desertec.org/downloads/summary_en.pdf viewed 17/3/08

327. Greenpeace International, European Renewable Energy Council (EREC), *Energy [R]evolution: A sustainable world energy outlook*, January 2007 p23

328. As any science fiction geek will tell you, terraforming means artificially engineering the natural systems of another planet to create an environment suitable for human life.

329. ETC Group, 'Gambling with Gaia', *Communique* Issue 93, January 2007 http://www.etcgroup.org/upload/publication/pdf_file/608

330. Stephen H. Schneider, 'Earth systems engineering and management', *Nature*, Vol 409, 18 January 2001

331. Michael Behar, 'How Earth-Scale Engineering Can Save the Planet,' *Popular Science*, June 2005. http://www.popsci.com

332. David Adam, 'US Government answer to global warming: Smoke and giant mirrors,' *The Guardian*, 27 January 2007.

333. Charles J. Hanley, 'Top Scientists Say Man May Need to Dirty Skies to Shield against Warming', *Associated Press*, November 16, 2006, as distributed by Environmental News Network

334. Discussion with negotiators at London Convention Meeting, 5 November 2007

335. Iron Fertilization News Blog, 'UK Environmental Minister: Ocean Fertilization Research Crucial', 28 February 2008, http://carbonsequestration.blogspot.com/ viewed 13/3/08

336. ETC Group, 'Gambling with Gaia', *Communique* Issue 93, January 2007 http://www.etcgroup.org/upload/publication/pdf_file/608

337. Wikipedia, 'Year without a summer', http://en.wikipedia.org/wiki/Year_Without_a_Summer viewed 12/2/08

338. Minnis et al., 'Radiative Climate Forcing by the Mount Pinatubo Eruption', *Science*, 5 March 1993

339. Paul J. Crutzen, 'Albedo Enhancement by Stratospheric Sulfur Injections: A Contribution to Resolve a Policy Dilemma?' *Climatic Change*, 77: p211–219, 2006

340. Ibid.

341. Alvia Gaskill, 'Aerosol Discussion', Global Albedo Enhancement Project, 22 October 2006 http://www.global-warming-geo-engineering.org/Aerosol-Discussion/

342. ibid.

343. Paul J. Crutzen, 'Albedo Enhancement by Stratospheric Sulfur Injections: A Contribution to Resolve a Policy Dilemma?' *Climatic Change* 77, 2006, 211–219

344. ibid.

345. IM Held, TL Delworth, J. Lu, KL Findell, and TR Knutson, 'Simulation of Sahel drought in the 20th and 21st centuries'. *Proceedings of the National Academy of Sciences, USA* vol. 102, no. 50, 13 December 2005, p17891-17896. DOI: 10.1073/pnas.0509057102.
Also; M Biasutti and A Giannini, 'Robust Sahel drying in response to late 20th century forcings'. *Geophysical Research Letters*, vol. 33, no. 11. DOI: 10.1029/2006GL026067. 8 June 2006.
Also; JE Kristjansson et al, 'Response of the climate system to aerosol direct and indirect forcing: Role of cloud feedbacks'. *Journal of Geophysical Research -Atmospheres* vol. 110, no. D24.23, December 2005

346. Alvia Gaskill, 'Introduction', *Global Albedo Enhancement Project*, http://www.global-warming-geo-engineering.org/Albedo-Enhancement/Introduction/ag1.html, viewed 12/02/08

347. Alvia Gaskill, 'Costs and Financing', *Global Albedo Enhancement Project*, http://www.global-warming-geo-engineering.org/Albedo-Enhancement/Surface-Albedo-Enhancement/Costs-Financing/ag41.html viewed 12/02/08

348. Alvia Gaskill, 'Disadvantages', *Global Albedo Enhancement Project*, http://www.global-warming-geo-engineering.org/Albedo-Enhancement/Introduction/Disadvantages/ag3.html viewed 12/02/08

349. Alvia Gaskill, 'Disadvantages', *Global Albedo Enhancement Project*, http://www.global-warming-geo-engineering.org/Albedo-Enhancement/Introduction/Disadvantages/ag3.html viewed 12/02/08

350. Alvia Gaskill, 'Advantages', *Global Albedo Enhancement Project*, http://www.global-warming-geo-engineering.org/Albedo-Enhancement/Introduction/Advantages/ag2.html viewed 12/02/08

351. R. Angel, 'Feasibility of cooling the earth with a cloud of small spacecraft near the inner Lagrange Point (L1).' *Proceedings of the National Academy of Sciences, USA*, doi:10.1073/pnas.0608163103.
Also; Bob Henson, 'Big Fixes for Climate?' *UCAR Quarterly*, National Center for Atmospheric Research, Fall, 2006. http://www.ucar.edu/communications/quarterly/fall06/bigfix.jsp viewed 12/2/08

352. Bob Henson, 'Big Fixes for Climate?' *UCAR Quarterly*, National Center for Atmospheric Research, Fall, 2006. http://www.ucar.edu/communications/quarterly/fall06/bigfix.jsp viewed 12/2/08

353. David Adam, 'Shiny crops could slow global warming, scientists say', *The Guardian*, 8 January 2008 (Peer reviewed article to be published later in 2008)

354. ibid.

355. Halfbakery blog, 'Ping Pong Albedo Enhancement', 24 May 2006 http://www.halfbakery.com/idea/Ping_20Pong_20Albedo_20Enhancement#1148492186

356. Royal Society, *Ocean acidification due to increasing atmospheric carbon dioxide*, June 2005. http://royalsociety.org/displaypagedoc.asp?id=13539

357. Royal Society, *Ocean acidification due to increasing atmospheric carbon dioxide*, June 2005. http://royalsociety.org/displaypagedoc.asp?id=13539

358. Virgin Earth Challenge website, http://www.virginearth.com/ viewed 12/2/08

359. ibid.

360. Proposed by Climos (www.climos.com) and Planktos (www.planktos.com)

361. Proposed by Ocean Nourishment Corporation (www.oceannourishment.com)

362. Proposed by Atmocean (www.atmocean.com)

363. Barker, T., I. Bashmakov, A. Alharthi, M. Amann, L. Cifuentes, J. Drexhage, M. Duan, O. Edenhofer, B. Flannery, M. Grubb, M. Hoogwijk,F. I. Ibitoye, C. J. Jepma, W.A. Pizer, K. Yamaji, 2007: 'Mitigation from a cross-sectoral perspective'. *Climate Change 2007: Mitigation. Contribution of Working Group III to the Fourth*

Assessment Report of the Intergovernmental Panel on Climate Change [B. Metz, O.R. Davidson, P.R. Bosch, R. Dave, L.A. Meyer (eds)], Cambridge University Press, Cambridge, UK and New York, NY, USA.

364. See for example: Michelle Allsopp, David Santillo and Paul Johnston, *A scientific critique of oceanic iron fertilization as a climate change mitigation strategy*, Greenpeace Research Laboratories Technical Note, September 2007 www.greenpeace.to/publications/iron_fertilisation_critique.pdf

365. Ken O. Buesseler et. al., 'Ocean Iron Fertilization—Moving Forward in a Sea of Uncertainty', *Science* Vol 319 11, January 2008

366. Planktos Press Release, 'Planktos Indefinitely Postpones Ocean Iron Fertilization Project', 13 February 2008, http://www.planktos.com/, viewed 13/3/08

367. International Maritime Organization Press Release, 'Large-scale fertilization operations not currently justified, say Parties to international treaties', IMO Briefing 40, 16 November 2007. http://www.imo.org/includes/blastData.asp/doc_id=8708/Press%20briefing%2016-11-07.doc

368. Cleantech.com, 'Climos hires Tetra Tech for environmental analysis', 28 March 2008, http://media.cleantech.com/2635/climos-hires-tetra-tech-for-environmental-analysis

369. Climos Press Release, 'Climos Closes on $3.5M Series A Financing with Braemar Energy Ventures', 5th March 2008 http://www.climos.com/releases/2008/35release.html, viewed 30/3/08

370. Earth2Tech, 'Q+A with Climos CEO Dan Whaley', http://earth2tech.com/2008/03/06/qa-climos-ceo-dan-whaley/ viewed 30/3/08

371. ibid.

372. Climos website, 'Elements of a code of conduct for ocean fertilisation projects', http://www.climos.com/standards/

373. Video of protest at London Convention meeting, http://www.youtube.com/watch?v=W7fTF863d2w

374. ETC Group, 'Backgrounder: Ocean Nourishment Corporation', 5 November 2007, http://www.etcgroup.org/en/materials/publications.html?pub_id=660

375. See for example Kevin Smith, *The Carbon Neutral Myth*, Transnational Institute, February 2007 http://www.carbontradewatch.org/pubs/carbon_neutral_myth.pdf

376. Climate Care press release, 'JPMorgan acquires Climate Care; A big challenge needs a big approach', 26 March 2008, http://www.climatecare.org/news/jpmorgan-acquires-climate-care-a-big-challenge-needs_2008_03_26/

377. Ning Zeng, 'Carbon sequestration via wood burial', *Carbon Balance and Management 2008, 3:1*, 3 January 2008

378. ibid.

379. Peter Read and Jonathan Lermit, 'Bio-Energy with Carbon Storage (BECS): a Sequential Decision Approach to the threat of Abrupt Climate Change', 2003 http://www.etsap.org/worksh_6_2003/2003P_read.pdf

380. ibid.

381. ibid.

382. Johannes Lehmann, 'A handful of carbon', *Nature* 447 p143-144, 10 May 2007

383. J. Lehmann, J. Gaunt, & M. Rondon, ' Mitigation Adaptation Strategies', *Global Change* 11, 403–427 (2006).

384. Johannes Lehmann, 'A handful of carbon', *Nature* 447 p143-144, 10 May 2007

385. ibid.

386. Peter Read and Jonathan Lermit, 'Bio-Energy with Carbon Storage (BECS): a Sequential Decision Approach to the threat of Abrupt Climate Change', 2003 http://www.etsap.org/worksh_6_2003/2003P_read.pdf

387. Molly Bentley, 'Synthetic trees could purify air', *BBC News,* 21 February 2003, viewed 21/4/08 http://news.bbc.co.uk/1/hi/sci/tech/2784227.stm

388. The name of the chemical used is secret. Global Research Technologies website, 'Frequently Asked Questions', http://www.grestech.com/faq.php viewed 19/01/08

389. Global Research Technologies website, 'Frequently Asked Questions', http://www.grestech.com/faq.php viewed 19/01/08

390. ibid.

391. ibid.

392. ibid.

393. Synthetic Genomics website,'The Science of Synthetic Genomics', http://www.syntheticgenomics.com/science.htm viewed 30/3/08

394. Johann Hari, 'The last green taboo: Engineering the planet', *The Independent*, 4 October 2007

395. J. Hansen, et al, 'Dangerous human-made interference with climate: a GISS model E study' *Atmos. Chem. Phys.* 7, 2287–2312, 2007

396. George Monbiot, *Heat: How to Stop the Planet Burning*, Penguin, 2006 p198-199

397. Jim Thomas, pers. comm.

398. 'Green and Drax', *Process Engineering*, 15 August 2007 http://www.processengineering.co.uk/Articles/301540/Green+and+Drax.htm

399. Marine Current Turbines press release, 'SeaGen completed: world's first megawatt-scale tidal turbine installed', 20 May 2008 http://www.marineturbines.com/3/news/article/9/seagen_completed__world_s_first_megawatt_scale_tidal_turbine_installed/

400. 'Tidal power farm plan off coast', *BBC News,* 7 February 2008, viewed 15/7/08 http://news.bbc.co.uk/1/hi/wales/north_west/7232738.stm

Carbon Capture and Storage diagram, page 36: Intergovernmental Panel on Climate Change, *Special Report on Carbon dioxide Capture and Storage*, Cambridge University Press, 2005, figure TS01